AUTO-BIOGRAPHY

This page enables you to compile a list of useful data on your car, so that whether you're ordering spares or just checking the tyre pressures, all the key information - the information that is 'personal' to your car - is easily within reach.

Registration number: ....

Model: ....

Body colour: ....

Paint code number: ....

Date of first registration: ....

Date of manufacture (if different): ....

VIN (or 'chassis') number: ....

Engine number: ....

Gearbox number: ....

Transfer box number (if applicable): ....

Front axle casing number (if applicable): ....

Rear axle casing number (if applicable): ....

Ignition key number: ....

Door lock key/s number/s: ....

Fuel locking cap key number (if fitted): ....

Alarm remote code (if fitted): ....

Alarm remote battery type: ....

Radio/cassette security code (if fitted): ....

Tyre size

    Front: .... Rear: ....

Tyre pressure (normally laden)

    Front: .... Rear: ....

Tyre pressure (fully laden)

    Front: .... Rear: ....

Insurance

    Name and address of insurer: ....

    ....

    Policy number: ....

Modifications

    Information that might be useful when you need to purchase parts: ....

    ....

    ....

Suppliers

    Address and telephone number of your garage and parts suppliers: ....

    ....

First published in 1995 by Porter Publishing Ltd.

Porter Publishing Ltd.
The Storehouse
Little Hereford Street
Bromyard
Hereford HR7 4DE
England

British Library Cataloguing in Publication Data.
A catalogue record for this book is available from the British Library.

ISBN 1-899238-09-3

Series Editor: Lindsay Porter
Design: Martin Driscoll, Lindsay Porter and Lyndsay Berryman
Layout and Typesetting: Unicorn Designs
Cover photography: Lindsay Porter
Printed in England by The Trinity Press, Worcester.

Every care has been taken to ensure that the material contained in this Service Guide is correct. However, no liability can be accepted by the authors or publishers for damage, loss, accidents, or injury resulting from any omissions or errors in the information given.

Titles in this Series:

Absolute Beginners' Service Guide
Caravan Owner's Manual & Service Guide
Classic 'Bike Service Guide
Ford Escort (Front Wheel Drive) & Orion Service Guide
Ford Fiesta (All models to 1995) Service Guide
Ford Sierra (All models) Service Guide
Land Rover Series I, II, III Service Guide
Land Rover Defender, 90 & 110 Service Guide

Metro (1980-1990) Service Guide
Mini (all models 1959-1994) Service Guide
MGB (including MGC, MGB GT V8 and
MG RV8) Service Guide
Vauxhall Astra & Belmont (All models-1995) Service Guide
Vauxhall Cavalier Service Guide
VW Beetle Service Guide
- With more titles in production -

Diesel Car Engines

Step-by-Step Service Guide

By Ivor Carroll

CONTENTS

Lindsay Porter
Porter Publishing Ltd

Introduction

Over the years, I have run any number of cars, from superb classic cars and modern cars, to those with one foot in the breakers yard. And I know only too well that any car is only enjoyable to own if it's safe, reliable and basically sound - and the only way of ensuring that it stays that way is to service it regularly. That's why we have set about creating this book, which aims to show the owner interested in diesel car servicing that there's nothing to fear; you really can do it yourself!

Making It Easy! Porter Publishing Service Guides are the first books to give you all the service information you might need, with step-by-step instructions, along with a complete Service History section for you to complete and fill in as you carry out regular maintenance on your car over the months ahead. Using the information contained in this book, you will be able to:

◆ see for yourself how to carry out every Service Job, from weekly and monthly checks, right up to longer-term maintenance items.

◆ understand how car diesel engines, in their various forms, actually work.

◆ enhance the value of your car by completing a full Service History of every maintenance job you carry out on your car.

I hope you enjoy keeping your car in trim while saving lots of money by servicing your diesel car's engine yourself, with the help of this book. Happy motoring!

Acknowledgements

No one organisation or person is to be thanked in particular for directly assisting me with the compilation and writing of this book! As hard as that sounds, this is the way things are when you're writing a generic service guide rather than one that's targeted at a particular car. I was out on a limb, gathering information from all manner of sources, and no one source in particular. Important companies which come foremost to mind, though, are Autodata Limited (technical publishers), Ford Motor Company, Rover Group, Vauxhall Motors and Volkswagen-Audi.

But if I were to give a really big thank you to anyone, it would be to my long-suffering wife, Sharon, who 'lost' me for a period of weeks while I immersed myself in my favourite subject - diesel cars! So big thanks to her, and also to Lindsay Porter, the Publisher, for giving me the opportunity to pull this book together.

As odd as it may sound, I am very much 'into' diesel cars. It's hard to remember just why, although my very first job, working for Peugeot Automobiles - great proponents of the diesel car cause - must have had something to do with it. Diesels make sense from virtually every viewpoint: they're inherently less harmful to the environment than petrol cars, they're more torquey, more durable, simpler than their petrol counterparts, they go an awful lot further on a tank of fuel... and they don't sputter to a short-circuited halt if you splash through a puddle!

In view of which, it would only be fair of me to give a big 'thanks' to Rudolf Diesel, who started the ball rolling all those years ago!

Ivor F. Carroll MIMI

SPECIAL THANKS

Thanks are also due to John Bishop of Bishop's Garage, Bromyard for assistance with the front cover picture - our good friends at Castrol for their continuing support and assistance - and the following companies for advice and use of line drawings: Robert Bosch Ltd, V L Churchill (Dieseltune), Ford Motor Company, LDV Limited, Lucas, Ricardo, Rover Group and V.A.G. (UK) Ltd.

OIL AND WATER DON'T MIX

It is important to remember that even a small quantity of oil is harmful to water and wildlife. And tipping oil down the drain is as good as tipping it into a river. Many drains are connected directly to a river or stream and pollution will occur.

Each year the National Rivers Authority deals with over 6,000 oil related water pollution incidents. Many of these are caused by the careless disposal of used oil.

The used oil from the sump of just one car can cover an area of water the size of two football pitches, cutting off the oxygen supply and harming swans, ducks, fish and other river life.

OIL POLLUTES WATER
USE YOUR BRAIN-
NOT THE DRAIN!

Follow the Oil Care Code

◆ *When you drain your engine oil - don't oil the drain!* Pouring oil down the drain will cause pollution. It is also an offence.

◆ Don't mix used oil with other materials, such as paint or solvents, because this makes recycling very difficult.

◆ Take used oil to an oil recycling bank. Telephone FREE on 0800 663366 to find the location of your nearest oil bank, or contact your local authority recycling officer.

OIL CARE
FOLLOW THE CODE

This book is produced in association with Castrol (U.K.) Ltd.

"Cars have become more and more sophisticated. But changing the oil and brake fluid, and similar jobs are as simple as they ever were. Castrol are pleased to be associated with this book because it gives us the opportunity to make life simpler for those who wish to service their own cars. Castrol have succeeded in making oil friendlier and kinder to the environment by removing harmful chlorine from our range of engine lubricants which in turn prolong the life of the catalytic converter (when fitted), by noticeably maintaining the engine at peak efficiency. In return, we ask you to be kinder to the environment too... by taking your used oil to your Local Authority Amenity Oil Bank. It can then be used as a heating fuel. Please do not poison it with thinners, paint, creosote or brake fluid because these render it useless and costly to dispose of."

Castrol (U.K.) Ltd.

CHAPTER 1 - SAFETY FIRST!

It is vitally important that you always take time to ensure that safety is the first consideration in any job you do. A slight lack of concentration, or a rush to finish the job quickly can often result in an accident, as can failure to follow a few simple precautions. Whereas skilled motor mechanics are trained in safe working practices you, the home mechanic, must find them out for yourself and act upon them.

Remember, accidents don't just happen, they are caused, and some of those causes are contained in the following list. Above all, ensure that whenever you work on your car you adopt a safety-minded approach at all times, and remain aware of the dangers that might be encountered.

Be sure to consult the suppliers of any materials and equipment you may use, and to obtain and read carefully any operating and health and safety instructions that may be available on packaging or from manufacturers and suppliers.

IMPORTANT POINTS

ALWAYS ensure that the vehicle is properly supported when raised off the ground. Don't work on, around, or underneath a raised vehicle unless axle stands are positioned under secure, load bearing underbody areas, or the vehicle is driven onto ramps.

NEVER start the engine unless the gearbox is in neutral and the hand brake is fully applied.

NEVER drain oil or automatic transmission fluid when the engine is hot. Allow time for it to cool sufficiently to avoid scalding you.

TAKE CARE when parking vehicles fitted with catalytic converters. The 'cat' reaches extremely high temperatures and any combustible materials under the car, such as long dry grass, could ignite.

NEVER run catalytic converter equipped vehicles without the exhaust system heat shields in place.

NEVER attempt to loosen or tighten nuts that require a lot of force to turn (e.g. a tight oil drain plug) with the vehicle raised, unless it is properly supported and in a safe condition. Wherever possible, initially slacken tight fastenings before raising the car off the ground.

TAKE CARE to avoid touching any engine or exhaust system component unless it is cool enough so as not to burn you.

ALWAYS keep antifreeze, brake and clutch fluid away from vehicle paintwork. Wash off any spills immediately.

NEVER siphon fuel, brake fluid or other such toxic liquids by mouth, or allow prolonged contact with your skin. There is an increasing awareness that they can damage your health. Best of all, use a suitable hand pump and wear gloves.

ALWAYS work in a well ventilated area and don't inhale dust - it may contain asbestos or other poisonous substances.

WIPE UP any spilt oil, grease or water off the floor immediately, before there is an accident.

MAKE SURE that spanners and all other tools are the right size for the job and are not likely to slip. Do not use metric spanners and sockets on imperial fastenings, or vice-versa: close enough is not good enough. Never try to 'double-up' spanners to gain more leverage.

SEEK HELP if you need to lift something heavy which may be beyond your capability.

ALWAYS ensure that the safe working load rating of any jacks, hoists or lifting gear used is sufficient for the job, and is used only as recommended by the manufacturer.

NEVER take risky short-cuts or rush to finish a job. Plan ahead and allow plenty of time.

BE meticulous and keep the work area tidy - you'll avoid frustration, work better and lose less.

KEEP children and animals right away from the work area and from unattended vehicles.

ALWAYS wear eye protection when working under the vehicle or using any power tools.

BEFORE undertaking dirty jobs or coming into contact with used engine oil or diesel fuel, use a barrier cream on your hands as a protection against skin damage and infection. Preferably, wear thin gloves, available from DIY outlets.

DON'T lean over, or work on, a running engine unless strictly necessary, and keep long hair and loose clothing well out of the way of moving mechanical parts. Note that it is theoretically possible for fluorescent strip-lighting to make an engine fan appear to be stationary - check! This is the sort of error that happens when you're dog tired and not thinking straight. So don't work on your car when you're overtired!

REMOVE your wrist watch, rings and all other jewellery before doing any work on the vehicle - especially the electrical system. Never rest tools on the car battery top - they could cause a short-circuit and explosion.

ALWAYS tell someone what you're doing and have them regularly check that all is well, especially when working alone on, or under, the vehicle.

ALWAYS seek specialist advice if you're in doubt about any job. The safety of your vehicle affects you, your passengers and other road users.

Fire

Only use flammable solvents in a well ventilated area, and disconnect the car battery earth lead to prevent accidental electrical sparks when working underbonnet, which could ignite solvent vapours.

Invest in a workshop-sized fire extinguisher. Choose the carbon dioxide type or preferably, dry powder, but never a water type extinguisher for workshop use. Water conducts

electricity and can make worse a fuel oil-based fire.

Fumes

In addition to the fire dangers described previously, vapour from many solvents, thinners, and adhesives is highly toxic and under certain conditions can lead to unconsciousness or even death, if inhaled. The risks are increased if such fluids are used in a confined space so always ensure adequate ventilation when handling materials of this nature. Treat all such substances with care, always read the instructions and follow them implicitly.

Always ensure that the car is outside the work place in open air if the engine is running. Diesel exhaust fumes contain poisonous carbon monoxide, even though it isn't present in as large quantities as in petrol exhaust fumes.

Never have the engine running with the car in the garage or in any enclosed space.

Inspection pits are another source of danger from the build-up of fumes. Never drain fuel or use solvents, thinners, adhesives or other toxic substances in an inspection pit as the extremely confined space allows the highly toxic fumes to concentrate. Running the engine with the vehicle over the pit can have the same results.

Cooling

In order to raise the boiling point of the coolant, to reduce the likelihood of it boiling over, the cooling system is kept under pressure when the engine is hot, or even warm. This means that depressurisation of a hot or warm system is likely to result in boiling over. The system is depressurised if any part of it is opened - namely, the radiator or catcher tank pressure cap. There is therefore a very real danger of scalding if the system is opened when under pressure, as boiling water will erupt from the opening and can scald your hand.

This is why it is advisable to open the system only when the engine is cool. However, if it must be opened when still warm, release the cap very gently and slowly, so as to relieve pressure without losing any significant amount of coolant. It will require patience to relieve all pressure by this method. Use a thick rag over the cap, to stifle any potential eruption.

Mains Electricity

Best of all, use rechargeable tools and a DC inspection lamp, powered from a remote 12V battery - both are much safer! However, if

you do use a mains-powered inspection lamp, power tool etc, ensure that the appliance is wired correctly to its plug, that where necessary it is properly earthed (grounded), and that the fuse is of the correct rating for the appliance concerned. Do not use any mains powered equipment in damp conditions or in the vicinity of fuel, fuel vapour or the vehicle battery.

Also, before using any mains powered electrical equipment, take one more simple precaution - use an RCD (Residual Current Device) circuit breaker. Then, if there is a short, the RCD circuit breaker minimises the risk of electrocution by instantly cutting the power supply. Buy one from any electrical store or DIY centre. RCDs fit simply into your electrical socket before plugging in your electrical equipment.

The Battery

Never cause a spark, smoke, or allow a naked light near the vehicle's battery, even in a well ventilated area. A certain amount of highly explosive hydrogen is given off as part of the normal charging process. Care should be taken to avoid sparking by switching off the power supply before charger leads are connected or disconnected. Battery terminals should be shielded, since a battery contains energy and a spark can be caused by any conductor which touches its terminals or exposed connecting straps.

Before working on the fuel or electrical systems, always disconnect the battery earth (ground) terminal.

When charging the battery from an external source, disconnect both battery leads before connecting the charger. If the battery is not of the 'sealed-for-life' type, loosen the filler plugs or remove the cover before charging. For best results the battery should be given a low rate 'trickle' charge overnight. Do not charge at an excessive rate or the battery may burst.

Always wear gloves and goggles when carrying or when topping up the battery. Even in diluted form (as it is in the battery) the acid electrolyte is extremely corrosive and must not be allowed to contact the eyes, skin or clothes.

If you wish to jump-start one car from the battery of another, be sure to connect the jump leads - always one by one - in the right order to avoid potential damage to electrical components. This is positive (red) to positive, and negative (black) either to the negative battery terminal or to a good earth point on the engine.

Run the engine of the donor car at a high idle while jump-starting, to avoid rapidly flattening the donor battery. When the car engine has been successfully started, disconnect the jump leads in the opposite order to connection, and do so only when both engines have been returned to a normal tickover, or the donor engine has been switched off.

Brakes and Asbestos

Whenever you work on the braking system mechanical components, or remove front or rear brake pads or shoes:

i) wear an efficient particle mask,

ii) wipe off all brake dust from the work area (never blow it off with compressed air) after applying a proprietary brand of spray-on brake cleaner,

iii) dispose of brake dust and discarded shoes or pads in a sealed plastic bag,

iv) wash hands thoroughly after you have finished working on the brakes and certainly before you eat or smoke,

v) replace shoes and pads only with asbestos-free shoes or pads. Note that asbestos brake dust can cause cancer if inhaled.

Obviously, a car's brakes are among its most important safety related items. Do not dismantle your car's brakes unless you are fully competent to do so. If you have not been trained in this work, but wish to carry out brake maintenance or repairs, it is strongly recommend that you have a garage or qualified mechanic check your work before using the car on the road.

Brake Fluid

Brake fluid absorbs moisture rapidly from the air and can become dangerous, resulting in brake failure. You should have your brake fluid tested at least once a year by a properly equipped garage with test equipment and you should change the fluid in accordance with your vehicle manufacturer's recommendations or as advised in this book. Always buy no more brake fluid than you need. Never store an opened pack. Dispose of the remainder at your Local Authority Waste Disposal Site, in the designated disposal unit, not with general waste or with waste oil.

Engine Oils

Take care and observe the following precautions when working with used engine oil. Apart from the obvious risk of scalding when draining the oil from a hot engine, there is the danger from contaminates that are contained in all used oil.

Always wear disposable plastic or rubber gloves when draining the oil from your engine.

i) Note that the drain plug and the oil are often hotter than you expect! Wear gloves if the plug is too hot to touch and keep your hand to one side so that you are not scalded by the spurt of oil as the plug comes away.

ii) There are very real health hazards associated with used engine oil, more so with used diesel engine oil, which contains particularly complex and dangerous chemical compounds. According to one expert source, "Prolonged and repeated contact may cause serious skin disorders, including dermatitis and cancer". Use a barrier cream on your hands and try not to get oil on them. Where practicable, wear gloves and wash your hands with hand cleaner soon after carrying out the work. Keep oil out of the reach of children.

iii) NEVER, EVER dispose of old engine oil into the ground or down a drain. In the UK, and in most EC countries, every local authority must provide a safe means of oil disposal. In the UK, try your local Environmental Health Department for advice on waste disposal facilities. See *page 7*.

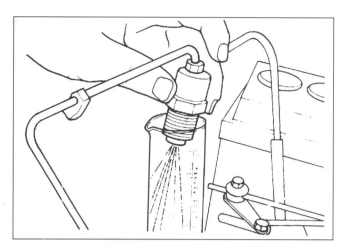

Fuel Injectors

The injectors are at the heart of engine performance, and it's likely that at some stage of ownership you are going to want to test them. Without specialised equipment, there is no way to check that an injector is doing its job properly without first removing it from the engine and running it connected to its supply pipe from the injection pump while you observe its

fuel spray. Unfortunately this is potentially very dangerous, as the fuel is injected at extremely high pressure - enough to pierce skin and enter the bloodstream. A dose of diesel fuel in the bloodstream is likely to be fatal, so you should always point an injector away from yourself and anyone else, and enclose its nozzle in a glass jar. Remember that as diesel fuel is harmful to skin, you should protect your hands before opening any part of the fuel system.

Plastic Materials

Work with plastic materials brings additional hazards into workshops. Many of the materials used (polymers, resins, adhesives and materials acting as catalysts and accelerators) readily produce very dangerous situations in the form of poisonous fumes, skin irritants, risk of fire and explosions. Do not allow resin or 2-pack adhesive hardener, or that supplied with filler or 2-pack stopper to come into contact with skin or eyes. Read carefully the safety notes supplied on the tin, tube or packaging.

Jack, Ramps and Axle Stands

Throughout this book you will see many references to the correct use of jacks, axle stands and similar equipment - and we make no apologies for being repetitive! This is one area where safety cannot be overstressed - your life could be at stake!

RAISING THE CAR BEFORE WORKING ON IT

Raising the Car - Safely!
You will often need to raise your car off the ground in order to carry out the Service Jobs shown here. To start off with, here's what you must never do: never work beneath a car held on a jack, not even a trolley jack. Quite a number of deaths have been caused by a car slipping off a jack while someone has been working beneath. On the other hand, the safest way is by raising a car on a proprietary brand of ramps. Sometimes, there is no alternative but to use axle stands. Please read all of the following information and act upon it!

When using car ramps:

(I) Make absolutely certain that the ramps are parallel to the wheels of the car and that the wheels are exactly central on each ramp.

Always have an assistant watch both sides of the car as you drive up. Drive up to the end 'stops' on the ramps but never over them!

Apply the hand brake firmly, put the car in first or reverse gear (or 'Park' in the case of an automatic).

SAFETY FIRST!

(II) Chock both wheels remaining on the ground, both in front and behind so that the car can't move in either direction. (A home-made wooden chock, cut at an angle so that it wedges under the tyre will be fine).

INSIDE INFORMATION: Wrap a strip of carpet into a loop around the first 'rung' of the ramps and drive over the doubled-up piece of carpet on the approach to the ramps. This prevents the ramps from skidding away, as they are inclined to do, as the car is driven on to them.

II

(III) On other occasions, you might need to raise the car with a trolley jack - invest in one if you don't already own one; the car's wheel changing jack is often too unstable. Place a piece of cloth over the head of the jack if your car is nicely finished on the underside. Ensure that the floor is sufficiently clear and smooth for the trolley jack wheels to roll as the car is raised and lowered, otherwise it could slip off the jack.

III

(IV) Look up in your owners' handbook or model-specific manual the safest places to position both jack and axle stands under your car.

IV

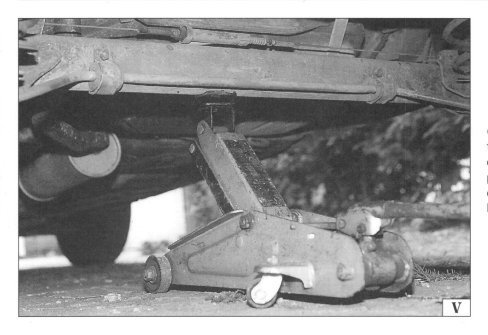

(V) Leave the jack in situ, but let the weight of the car rest firmly on the axle stands (note the plural, it's best to use more than one when you're raising the car like this).

(VI) If, for any reason, you can't use the location points recommended, take care to locate the top of the axle stand on a strong, level, stable part of the car's underside. Never use a movable suspension part because the part can move and allow the axle stand to slip, nor use the floor of the car, which is just too weak.

Just as when using ramps - only even more importantly! - apply the handbrake firmly once the car is fully raised by the jack (but NOT before!), put the car in first or reverse gear (or 'Park', in the case of an automatic) and chock both wheels remaining on the ground, both in front and behind.

Be especially careful when applying force to a spanner or when pulling hard on anything, when the car is supported off the ground. It is all too easy to move the car so far that it topples off the axle stand or stands. And remember that if a car falls on you, YOU COULD BE KILLED!

Caution is needed when working on the car whilst it is supported on an axle stand or a pair of axle stands. These are inherently less stable than ramps, so you must take much greater care when working beneath them. In particular:

- ensure that the axle stand is on flat, stable ground, never on ground where one side can sink in to the ground.

- ensure that the car is on level ground and that the handbrake is off and the transmission in neutral.

Whenever working beneath a car, have someone primed to keep an eye on you! If someone pops out to see how you are getting on every quarter of an hour or so, it could be enough to save your life!

Do remember that, in general, a car will be more stable when only one wheel is removed than if two wheels are removed in conjunction with two axle stands. You are strongly advised not to work under the car with all four wheels off the ground, on four axle stands. The car could then be very unstable and dangerous to work beneath.

When lowering the car to the ground, remember first to remove the chocks, release the handbrake and place the transmission in neutral.

FLUOROELASTOMERS

MOST IMPORTANT! PLEASE READ THIS SECTION!

If you service your car in the normal way, none of the following may be relevant to you. Unless, for example, you encounter a car which has been on fire (even in a localised area), subject to heat in, say, a crash-damage repairer's shop or vehicle breaker's yard, or if any second-hand parts have been heated in any of these ways.

Many synthetic rubber-like materials used in motor cars contain a substance called fluorine. These materials are known as fluoroelastomers and are commonly used for oil seals, wiring and cabling, bearing surfaces, gaskets, diaphragms, hoses and 'O' rings. If they are subjected to temperatures greater than 315 degrees C, they will decompose and can be potentially hazardous. Fluoroelastomer materials will show physical signs of decomposition under such conditions in the form of charring of black sticky masses. Some decomposition may occur at temperatures above 200 degrees C, and it is obvious that when a car has been in a fire or has been dismantled with the assistance of a cutting torch or blow torch, the fluoroelastomers can decompose in the manner indicated above.

In the presence of any water or humidity, including atmospheric moisture, the by-products caused by the fluoroelastomers being heated can be extremely dangerous. According to the Health and Safety Executive, "Skin contact with this liquid or decomposition residues can cause painful and penetrating burns. Permanent irreversible skin and tissue damage can occur". Damage can also be caused to eyes or by the inhalation of fumes created as fluoroelastomers are burned or heated.

After fires or exposure to high temperatures observe the following precautions:

1. Do not touch blackened or charred seals or equipment.

2. Allow all burnt or decomposed fluoroelastomer materials to cool down before inspection, investigations, tear-down or removal.

3. Preferably, don't handle parts containing decomposed fluoroelastomers, but if you must, wear goggles and PVC (polyvinyl chloride) or neoprene protective gloves whilst doing so. Never handle such parts unless they are completely cool.

4. Contaminated parts, residues, materials and clothing, including protective clothing and gloves, should be disposed of by an approved contractor to landfill or by incineration

according to national or local regulations. Oil seals, gaskets and 'O' rings, along with contaminated material, must not be burned locally.

WORKSHOP SAFETY - SUMMARY

1. Always have a fire extinguisher of the correct type at arm's length when working on the fuel system - under the car, or under the bonnet.

If you do have a fire, DON'T PANIC. Use the extinguisher effectively by directing it at the base of the fire.

2. NEVER use a naked flame near petrol or anywhere in the workplace.

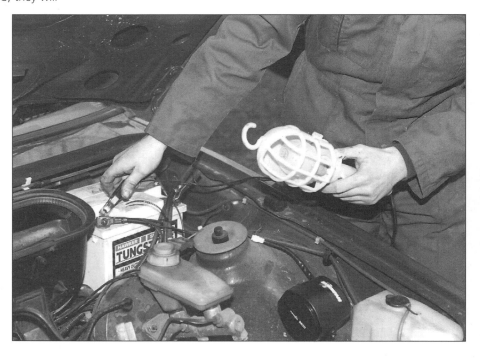

3. NEVER use petrol (gasoline) to clean parts. Use paraffin (kerosene) or white (mineral) spirits, but do not clean rubber components such as oil seals with it, as they may perish.

4. NO SMOKING! There's a risk of fire or transferring dangerous substances to your mouth and, in any case, ash falling into mechanical components is to be avoided!

5. BE METHODICAL in everything you do, use common sense, and think of safety at all times.

CHAPTER 2 - BUYING SPARES

Diesels have a well-deserved reputation for reliability, but there will, of course, be occasions when you need to buy spares in order to keep your pride and joy running. There are a number of sources of supply of the components necessary when servicing the car, the price and quality varying between suppliers. As with most things in life, cheapest is not necessarily best - as a general rule our advice is to put quality before price - this policy usually works out less expensive in the long run!

1. The obvious place to look for reference numbers on your diesel car is the under-bonnet area. The VIN (Vehicle Identification Number) plate is located at the bulkhead, the bonnet landing panel (as here)...

2. ...or on the inner wings on nearly all modern cars.

The engine and chassis numbers should match up with the DVLC V5 form (the 'logbook'). On some modern cars the chassis number is displayed in one lower corner of the windscreen.

Number (or VIN) and engine numbers. These can be helpful where parts changed during production, and can be the key to a more helpful approach by some parts salespeople! You may, by now have entered this key information on the Auto-Biography pages at the front of this book, for ease of reference. The photos on these pages show typical identification number locations.

Parts Factors/Motor Accessory Shops

Local parts factors and motor accessory shops can be extremely useful for obtaining servicing parts at short notice - many 'accessory' outlets open late in the evening, and on both days at weekends. However, as they tend to concentrate on 'mainstream' models, you will have to be very specific with details if yours is one of the scarcer diesels on UK roads. Remember that diesel cars are still in the minority.

Don't overlook the 'trade' motor factors outlets in the UK. Many of them have branches all over the country - you can find them in the Yellow Pages. Large motor factors chains such as these tend only to stock replacement parts of high quality, and at prices that more often than not undercut main dealers. If they don't have the parts you need in stock, they can usually obtain them within 24 hours. And remember that these outlets do sell to the public as well as to the 'trade'.

Buying at the Counter

If you're buying spares 'in person' rather than by post, try to avoid Saturday and Sunday mornings when buying - weekends are often very busy for parts counters, and you may find the staff have more time to help you if you are able to visit during the weekday, or in the evening, while on your way from work. At these times you are also less likely to have to queue for a long time!

INSIDE INFORMATION: Whenever possible, take the old part with you when going to collect the new replacement. And before you leave the premises, make a visual check - open the packaging if necessary - that the new part matches the old. If it doesn't - and you'll be surprised how often it doesn't! - at least you'll be saved a journey.

SOURCES

Main Dealers

How can you identify 'quality'? It's sometimes difficult, but parts from your main dealer are certain to be of the best quality, and, of course, they will be 'genuine' items with a comprehensive warranty (sometimes this is much longer than normal). In addition, the parts counter staff will be familiar with your vehicle, and should be only too pleased to help enthusiasts locate the spares required by means of a microfiche system and a computer linked to other dealerships of the same franchise.

Prices are sometimes reduced from the usual retail level - watch for special offers which are often listed at the parts counter but more often than not the main dealer is the most expensive source of spares.

When buying spares, have your diesel car's 'personal' details to hand - the date of registration and its Vehicle Identification

Buying Spares From Specialists

The number of diesel car specialist suppliers is on the increase, and at least one diesel fuel injection workshop with parts counter can be found in every major town. French and German car specialists abound - most of whom have excellent diesel expertise because of the strong presence of diesel cars from these countries.

The monthly magazine Diesel Car regularly carries adverts for these companies, and these list the most popular spares and prices. Some offer a choice of quality on many items. These will usually carry a note to the effect that they are O/E (original equipment) quality, or otherwise, and usually the country of origin. It's true to say that some 'pattern' parts are just about as good as the originals; it's equally true to say that some certainly aren't! When it comes to fuel and oil filters, don't be tempted to buy replacements of spurious manufacture. Whether you choose O/E or pattern for other spares will depend on the depth of your wallet and whether or not you're intent on staying 'original'. However, we would always recommend buying the best quality braking, steering and suspension products you can, regardless of price, because your life depends on them.

Ordering Spares by Post

Most specialists offer a postal service, with payment being by credit card. The points to check are a) the cost of carriage, b) whether VAT is included and c) if there is a cut-off point where you don't pay carriage, or where it is cheaper. For example, it could be that orders over £20 are not subject to a £5 carriage charge. If your order comes to £18, you'd be better off adding a spare alternator belt or filter to it, rather than pay £5 for 'nothing'.

Buying Secondhand

This can be a very expedient thing to do when a new O/E replacement part is exorbitantly expensive, or difficult or impossible to locate. Breakers' yards are the obvious venue for buying secondhand spares, and so are owner's clubs, which you may find for ubiquitous diesel cars such as the Citroen BX and VW Golf. Naturally, you need consider very seriously what it is you're buying. Purchasing any safety items - braking, steering, suspension - without being absolutely sure of their provenance is dangerous indeed, and not something we would recommend. Even if you're buying an unused, but old part from someone, the odds are it has been standing for some years, and may be distorted, internally corroded or perished (obviously depending on the component type). Make sure that safety-related parts are still serviceable and are not, for example, covered in a fine coating of rust.

That's not to decry buying secondhand altogether. Buying, say, a fuel injection pump or alternator which you know to be 'low mileage' units, to replace your worn out components makes great sense. Equally, and moving on from the purely 'diesel' parts of your car, trim panels and other interior parts can often be obtained at a fraction of the new cost with no risk to safety.

Clubs

Owner's clubs set up for diesels are rare, but there are clubs in this country catering for Citroens, Peugeots and VWs, which number diesel fanatics and specialists among their ranks. Most will be able to offer general spares information while the bigger ones often have a specialist tool hire service, to save you buying expensive tools you'll only use once every blue moon. If it is appropriate to the diesel car you drive, it's worth joining an enthusiastic, established club and make the best of the benefits they offer.

Fuel System Parts

Fuel injection pump

3. These are the components with which you are most likely to become involved as you nurture your diesel car through life. The injection pump is the heart of the fuel injection system, and the most complicated and costly part. Fortunately it rarely goes wrong, and it usually outlasts the engine. However, poor fuel filtration can spell its untimely demise, and then there is no choice but to buy a replacement unit from a main dealer or a Lucas or Bosch diesel specialist. This is a Bosch pump, and Bosch specialists tend to be the better value option, especially as they may be able to provide quality reconditioned units.

4. It isn't sufficient to ask for a replacement pump for a particular model of car, as the exact engine type and, sometimes even the year and month of engine manufacture can dictate pump differences. Replacement pumps must be matched with the identification numbers stamped on a plate on the original pump body. Fuel injection pumps are NOT to be opened up or tampered with by DIYers! This is an example of a Lucas unit.

Fuel Injectors

If all goes well, a replacement pump will never be an issue. Replacement injectors may well be, however. It's generally good advice to renew or overhaul injectors at about 70,000 miles to ensure continued efficient engine operation. Replacement injector units are notoriously expensive, and the mistake many people make is to buy them unnecessarily, when all that is needed is a new or reconditioned injector nozzle assembly. Diesel injection specialists can overhaul injectors for you at very reasonable prices.

Glow plugs

5. These are consumable items, and in the worst case, you may have to replace them as complete sets as often as you would spark plugs on a petrol engine. There is no hard and fast rule to glow plug longevity, but when they are due for replacement, you will find spares available for popular models at every diesel specialist's premises. Again, these should be matched by stamped markings, (arrowed) although it's not vital to buy O/E manufacture, as all manufacturers of glow plugs are specialists in the field. (Illustration, courtesy Peugeot)

Checks on Running Gear Components

This manual may be about diesel-specific matters, but where safety is concerned we feel it's always worth making reminders. So always take great care when purchasing 'hardware' for the steering, suspension and braking systems, which are obviously vital for safety. Although many outlets sell 'reconditioned' components, particularly engines and transmissions, on an 'exchange' basis, the quality of workmanship and the extent of the work carried out on such units can vary greatly. Therefore, if buying a rebuilt unit, always check particularly carefully when buying. It has to be said that, wherever possible, reconditioned units are best obtained from main dealers, major national retail outlets, or from reputable specialist suppliers. Always talk to fellow owners before buying - they may be able to direct you to a supplier offering sound parts at reasonable prices. When buying, always enquire about the terms of the guarantee (if any!).

In any event, the following notes should help you make basic checks on some of the commonly required components:

Before Fitting...

Faster moving consumer items such as alternators, starter motors, clutch assemblies, radiators and brake vacuum pumps may be available as exchange items rather than outright purchases - check beforehand. Ensure that you fully rotate the operating shaft of any replacement alternator, pump or motor that you have bought, feeling for any undue free play, roughness, stiffness, or 'notchiness' as you do so. Reject any units showing signs of any of these problems. It is better to find out that something is wrong before you waste time fitting it!

Suspension, Brakes & Tyres

Again, safety related so worthy of mention. We would strongly advise against buying secondhand brake, suspension, and steering components, unless you know the source of the parts, and really are sure that they are in first class condition. Even then, be sure that you see the vehicle they have been taken from, and avoid any such parts from accident-damaged cars.

In every case, ensure that the components you are buying are compatible with your particular vehicle, and carry out basic checks to ensure that they too are not badly worn.

Shock absorbers aren't worth buying on a used basis - standard units are rarely expensive, especially not if bought from motor factors. Don't forget that shock absorbers should always be replaced in pairs and preferably in complete sets of four.

6. Similarly, brake pads and shoes, even replacement brake discs and drums. Always renew these items in complete axle sets to keep braking even and properly balanced. In general, consumable parts of all types will be cheapest from a high street motorists' parts store - so shop around.

For the ultimate in long life, roadholding and wet grip, brand new radial tyres from a reputable manufacturer offer the best solution. Buy good quality tyres, look after them well, and they'll last many years.

Remoulds are available at lower initial cost, but life expectancy is not as long as with new tyres, so they tend to be only a short-term economy.

Secondhand tyre outlets are becoming increasingly common. However, if you purchase such tyres, you are taking a risk in that you have no knowledge of the history of the tyres or what has happened to them, how they have been repaired, and so on. A report conducted by the RAC revealed that a very high percentage of second-hand tyres in their test sample had very dangerous faults, such as damaged walling. Their advice, and we would agree, is to stick to top quality, brand new tyres from a reputable manufacturer. They may cost a little more, but at least you will have peace of mind, and should be able to rely on their performance in all road and weather situations. After all, your life - and those of other road users - could depend on it!

Conclusion

Finally, if you want to buy quality and save money, you must be prepared to shop around. Ring each of your chosen suppliers with a shopping list to hand and your car's personal data from the Auto-Biography at the front of this book in front of you. Keep a written note of prices - including VAT, delivery etc., whether the parts are proper 'brand name' parts or not and - most importantly! - whether or not the parts you want are in stock. Parts expected 'soon' have been known never to materialise.

CHAPTER 3 - HOW IT WORKS

Most people can recognise a diesel engine from its clatter on tick-over and the - shall we say, distinctive - exhaust pipe output! But ask them what makes it tick (or clatter, as the case may be) and it's not so easy. In this chapter, we aim to explain what it is that makes a diesel engine work, as well as some of the advantages of Professor Diesel's brainchild over the petrol engine.

What is a Diesel Engine?

1. A car's diesel engine is very similar in construction to a petrol engine, and it operates on the same four-stroke cycle. The principal difference between it and the petrol engine is that it doesn't need an ignition system (spark plugs and coil, etc.) to fire it. Instead, intake air is compressed by the piston of each cylinder's to the point where it becomes hot enough to ignite fuel.

2. You may be surprised that the simple act of compressing air can make it so hot, but it's certainly no secret to cyclists that the act of compressing air heats it up, which is why many bicycle tyre pumps can become too hot too handle!

But back to the diesel engine: When the piston is at the top of its compression stroke, and the air is hot (about 700 degrees C) diesel fuel is injected into the cylinder, whereupon it is ignited by the heat in the air. The diesel fuel-in-air ignites and expands with almost explosive power, forcing the piston back down the cylinder and helping to power the engine, the car, and enough 'shove' to compress the next charge of air right up to 700 degrees C again. It's a wonderfully simple system, and it does away with inherently unreliable and complicated ignition electrics.

But apart from the elimination of spark plugs, the diesel engine has other advantages. The greatest of these is that by compressing its intake air to a much greater extent than a petrol engine compresses its charge (a typical compression ratio of 22:1 against 9:1), the diesel engine makes itself more "thermally efficient". This means that it more efficiently produces power from a given quantity of fuel. The result: many more miles-per-gallon than its petrol-fuelled equivalent.

Diesel History

It was in 1890 that Doctor Rudolf Diesel developed the theories of the "economical thermal motor" which, through its high cylinder compression, significantly improved efficiency. Although he was the first to file a patent for this "compression-ignition" engine, an engineer named Ackroyd Stuart previously had similar ideas. He proposed an engine in which air was drawn into the cylinder, compressed, then forced - at the end of the compression stroke - into a bulb into which fuel was sprayed. To start the engine, the bulb was heated externally by a blow lamp, and once running it kept itself going without external heat.

Ackroyd Stuart wasn't concerned with the advantages of operating very high compression pressures, he was simply experimenting with the possibility of avoiding the use of spark plugs, so he missed out on the greatest advantage - fuel efficiency. This is why the term "diesel engine" has remained, as it is Dr. Diesel's theories which form the basis of the modern "compression-ignition" engine.

The Cycle of Operations

This is the full four-stroke cycle of diesel engine operation.

3A. On the first (downward, induction) stroke of the piston a fresh charge of air alone is drawn into the cylinder through the open inlet valve.

3B. On the second (upward, compression) stroke, all valves

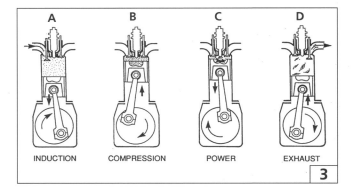

A	B	C	D
INDUCTION	COMPRESSION	POWER	EXHAUST

are closed, and this air is compressed into a volume typically 22 times smaller than the total cylinder volume, so becoming very hot.

3C. Just before the start of the third (downward, power) stroke, diesel fuel is injected into the combustion chamber through a nozzle. During injection the fuel is split up into tiny particles which mix evenly with the compressed air to make a combustible, self-igniting charge. Energy is unleashed by combustion as the piston starts its descent on the power stroke. Injection continues, causing the burning fuel to maintain constant pressure on the piston.

3D. The exhaust valve opens as the piston starts its fourth (upward, exhaust) stroke, evacuating exhaust gases through the valve as it travels.

Advantages & Disadvantages

4A

4B

4A. Early diesel engines were noisy, sluggish, rough and smoky, such as this one used in this 1935 Citroen Rosalie, one of the world's first production diesel cars, fitted with a British Ricardo diesel engine. (Illustration, courtesy Ricardo)

4B. The engine layout is remarkably similar to those of today's cars. (Illustration, courtesy Ricardo)

5A

5B

5A. Today's light diesels have been so extensively refined however, that it is often very difficult to tell whether you're driving a diesel or a petrol... sometimes only the greater flexibility of the diesel engine (or an open window) give the game away. This is a 1995 Rover 620 SLDi with 2.0 litre diesel engine.

5B. The modern engine's 'clothes' make it look very different from its forebear although differences *are* superficial under the skin.

Despite its widespread use in cars, the petrol engine is comparatively inefficient, capable of converting only about 26% of the energy in its fuel as useful work. The diesel, though, typically returns a 36% fuel-efficiency, and so is usually more economical to run. As well as other obvious advantages, such as the elimination of an ignition system, the diesel provides high torque (pulling power) over most of its speed range, making it much more flexible to drive. This in turn means that the vehicle it is powering can be equipped with higher gearing to provide more relaxed cruising. Although it is true that petrol engines are higher-revving than diesels, because of the diesel's limitations of mechanical fuel-injection and heavy internal componentry, in practice most drivers find flexibility to be far more useful than hard-revving power.

Diesel engines also provide better engine-braking than do petrol engines. In other words, there is more of a tendency for a diesel car to slow if you lift off the throttle, as the high cylinder compression provides rotational resistance, so the engine acts as a brake on the car's wheels. This is useful for giving the diesel car a feeling of greater stability and control downhill, and for the same reason it is a real asset to have a diesel engine fitted to an off-road vehicle.

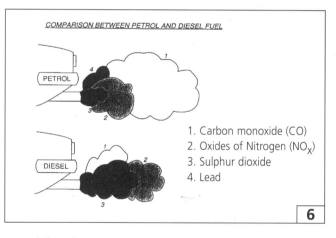

COMPARISON BETWEEN PETROL AND DIESEL FUEL

PETROL

DIESEL

1. Carbon monoxide (CO)
2. Oxides of Nitrogen (NO_x)
3. Sulphur dioxide
4. Lead

6

6. And the advantages don't stop there. The diesel's exhaust gases are relatively harmless compared with those of the petrol engine. The most harmful ingredient, carbon monoxide (CO), is virtually non-existent in the diesel's emissions, so the only undesirable gases present in any significant quantities are hydrocarbons (HC), (not shown in the above chart) oxides of nitrogen (NOx shown as N above), and soot in the form of black smoke. Black smoke (made up of soot, or particulates) has been tenuously linked to asthma and lung cancer - although the greatest offenders here are the heavy commercial diesels, such as trucks and buses, which are often old and poorly maintained.

HOW IT WORKS

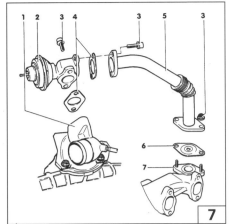

7

7. Even these gases are being successfully tackled by modern technology: NOx can be largely reduced by Exhaust Gas Recirculation (EGR). Exhaust gas recirculation takes some exhaust from the exhaust manifold (7) via duct (5) to intake manifold (1). The process is controlled by valve (2) and it reduces combustion temperature, and so the production of NOx is reduced. (Illustration, courtesy V.A.G.)

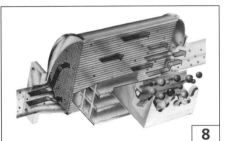

8

8. Oxidation-type catalytic converters are in use to substantially reduce hydrocarbons and the remaining CO. And as for the soot, improvements in

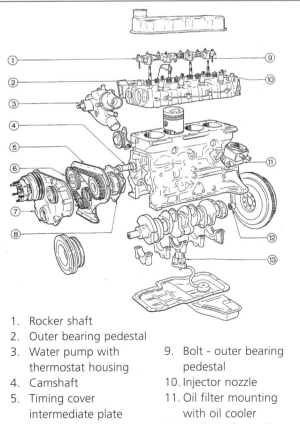

1. Rocker shaft
2. Outer bearing pedestal
3. Water pump with thermostat housing
4. Camshaft
5. Timing cover intermediate plate
6. Timing chain
7. Timing cover
8. Injection pump gear mounting flange
9. Bolt - outer bearing pedestal
10. Injector nozzle
11. Oil filter mounting with oil cooler
12. Crankshaft needle roller bearing
13. Oil pump with intake pipe

9

fuel injection and combustion are rapidly eliminating it. Good, regular maintenance of diesel engines helps keep black smoke down to a minimum. This is a cutaway of the diesel's oxidation catalytic converter, which substantially reduces the emissions of CO and hydrocarbons and the attendant nasty 'diesel smell'. (Illustration, courtesy V.A.G.)

One other important aspect - and it's safety-related - is that diesel fuel isn't volatile (it doesn't vaporise easily) so the likelihood of vehicle fires is much smaller with diesels - particularly as there is no ignition system to start them!

Of course there are disadvantages, among them the clattering noise from which most diesels suffer when ticking over, and the unpleasant greasiness of the fuel. But these really are small penalties to set alongside the substantial gains.

Diesel Engine Construction

9. The diesel engine's basic construction is virtually the same as that of the petrol engine. This exploded view of the Ford/Peugeot 2.3 litre diesel engine shows how similar construction is. The cylinder-block and bottom-end (i.e.crank-shaft assembly) are the same, as are the connecting rods and pistons. (Illustration, courtesy Ford Europe Ltd)

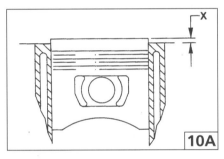

10A

10A. The piston crowns, however, are specifically designed to suit diesel combustion, and often, because of the very high compression, the piston tops are higher than the top face of the cylinder block when each piston is at the top of its stroke. (Illustration, courtesy Ford Europe Ltd)

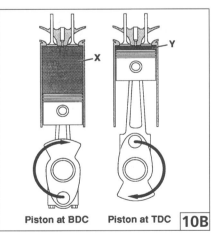

Piston at BDC   Piston at TDC   **10B**

10B. The compression ratio of an engine is the ratio between the volume 'X' above the piston when it is at the bottom of its stroke (BDC) compared with the volume 'Y' when it is at the top of its stoke (TDC). (Illustration, courtesy V L Churchill/LDV Limited)

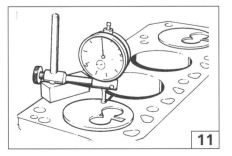

11

11. When diesel engines are rebuilt, the piston protrusion is measured with a dial gauge before a suitable head gasket is selected. (Illustration, courtesy Ford Europe Ltd)

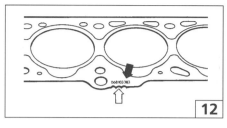

12. Selective thickness head gaskets are usually identified by different numbers of notches on the edge. The correct thickness will ensure that the piston and valves don't collide when the engine is running.

13. The valvegear is also conventional, as is the camshaft drive arrangement, with the exception that it also drives a fuel injection pump.

The major differences lie in the air intake system, which has no throttle butterfly, the design of the cylinder head and, of course, the presence of an injection pump in place of an ignition distributor and a petrol engine's carburettor or fuel injection system.

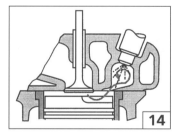

14. Most diesel car engines are of the indirect injection type, where combustion starts in a pre-combustion ante-chamber. Note the swirling of the compressed intake air. The pre-combustion chamber, one for each cylinder, is positioned in the cylinder head, with an injector and a glow plug protruding into it. Not all diesel engines have pre-combustion chambers, as there are two fundamental types of diesel engine in cars - indirect injection (IDI), and direct-injection (DI) - and the latter doesn't use ante-chambers. (Illustration, courtesy V L Churchill/LDV Limited)

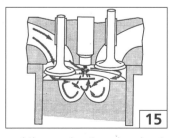

15. If you own a Maestro or Montego Diesel, or a VW or Audi bearing a 'TDI' badge, you have the direct injection type; virtually every other diesel car is an IDI. The more efficient direct injection diesel has no ante-chamber, and the combustion chamber is formed in the piston crown. Note the important swirl that is imparted to incoming air by the design of the inlet port. (Illustration, courtesy V L Churchill/LDV Limited)

If you want to know more about the reasons why these two types exist, and their exact differences, you'll find an explanation under the *INJECTORS* section later in this chapter.

Diesel Engine Lubricating Oil

There are engine oils on the market that are specific to diesel engines. There are also multigrade oils which are suitable for use in both diesel and petrol engines.

The way to tell whether a given oil is suitable for use in your diesel engine is to look at the API or CCMC specification quoted for it on the pack. More of which shortly; first we'll look at the ingredients of modern motor oils and the differences in lubrication requirements between diesel and petrol engines.

If we consider only mineral oils for the time being, we have much of a muchness between one oil intended for diesels and another for petrols, before the carefully formulated additive packages are 'stirred in'. The 'package', whether applied to a diesel or a petrol engine application, comprises a basic group of important additives; the proportions and exact nature of each are likely to alter depending upon the application. This group incorporates the following ingredients:-

a) 'Detergents': usually metallic compounds that help keep engine components clean and prevent deposits from forming, particularly on pistons and rings. They control the problem of 'varnish' formation, which is a coating created from oil oxidation products that bake on to components during high temperature use.

b) 'Dispersants': non-metallic compounds often used in conjunction with detergents. Dispersants are vital for keeping soot and other undesirable deposits from settling out onto engine surfaces - they keep the contaminants harmlessly in suspension. 'Sludge' - a black, treacle-like emulsion of water, combustion products and oil formed under cold-running conditions, is also combated by dispersants.

c) 'Anti-oxidants': prevent the oil from degrading with increasing temperature, and so lengthen its service life.

d) 'Anti-wear agents': particularly important to minimise wear on rubbing surfaces such as those of piston/cylinder, cam/follower and plain bearings.

e) 'Corrosion inhibitors': protect lubricated metal surfaces against chemical attack by water, which causes rust, and acids which may form as oxidation products in a deteriorating oil, or may be introduced into the oil by by-products of combustion. Corrosion naturally leads to wear.

f) 'Anti-foam additives': needed because omnipresent air in the engine crankcase is whipped into the oil by the movement of oil and components. Aerated oil cannot perform properly as a lubricant and so aeration must be suppressed.

g) 'Pour point depressant': engine oil needs to remain reasonably thin when cold so that it can circulate properly and not impose too much of a load on the engine's starting system; the depressant is an additive that lowers the pour point.

h) Viscosity index (VI) improver: this reduces the tendency of an oil to change its viscosity (thickness) as temperature varies. A multigrade oil is one containing VI improvers that make it suitable for use over a much wider range of temperatures than a single-grade oil. Wide temperature range use is a requirement of car engines, whether diesel or petrol, since the range of temperatures that lies between a cold start on a winter morning and a long motorway thrash on a hot day is vast. The internationally-accepted SAE multigrade viscosity classifications comprise primarily a cold viscosity rating, followed by a 'W' for 'winter' (ie. '10W'), and secondarily the hot viscosity rating of the oil - say '40'. The higher the number, the less viscous, or thinner, is the oil. Today's popular multi-grade for petrol engines of 10W 40 is thinner under both cold and hot usage conditions than was the staple 20W 50 multigrade of a decade ago. Engine oil thickness has been gradually reduced over the years from 20W 50, through 15W 40 and currently to 10W 30 on some modern petrol and diesel engines - a change that has been made possible by better designed engines with reduced internal friction and much improved oil control.

Which Multigrade Rating is Best?

The advantages of using a thinner engine oil are many, and include better fuel returns due to decreased friction, easier, more reliable cold starting for much the same reason, and lighter, smaller and cheaper starter motors and batteries.

16. The currently preferred multigrade viscosity for modern car diesel engines is 10W 40. Whilst an engine will have a particular oil grade specified for it by its manufacturer, the grade chosen will be ideal for the engine's new, perfect, state, but not necessarily ideal later on in life. In some cases where an engine is displaying certain signs of 'old age', such as bearing noise on start-up, or low compression, or even in cases where an engine is fine but has covered a reasonably high mileage - maybe 100,000 or more miles - it may prove advantageous to use a thicker oil like a 20W 50, since the greater viscosity can help restore cylinder compressions and affords greater protection to sloppy bearings.

The differences between petrol and diesel-specific engine oils are actually relatively few, the obvious differences in lubrication requirements being as follows.

Why Diesel Engine Oil is Different

Where petrol has a sulphur content in the region of 0.03%, diesel fuel, at about 0.2%, has a considerably higher content. Sulphur is pretty undesirable stuff in a fuel because its combustion produces corrosive sulphuric acid and environmentally nasty sulphur oxides in the exhaust. Because of these higher levels, diesel lubricating oil has more harmful acids dumped in it, and therefore requires more detergents and an inherent surplus of alkalinity to help neutralise the ill-effects of acids. Incidentally, this production of acids is one of the major factors forcing the frequent oil change requirements of the diesel. Another is dilution of the lubricating oil by fuel, which in the case of the petrol engine evaporates from the oil, something which diesel fuel cannot readily do.

The diesel's production of soot is the other major factor determining a difference between lubricating requirements. The diesel tends to burn 'sooty', so greater quantities of carbon deposits find their way into the engine oil and out of the exhaust pipe in the form of the much-lamented 'particulates'. This unwanted internal effect must be combated by a different and stronger cocktail of dispersive agents within the oil, compared with that of petrol engine oil. The excess carbon held in suspension in the diesel's oil can build up in time to thicken it, in direct opposition to the effects of unburned fuel which thins it. And the oil's detergents are not lasting, but get used up just like we consume our own detergents in the bathroom, so eventually old oil can be depleted of these vital agents. As you can see therefore, the oil's design properties and additive balances can soon be lost - hence the importance of changing engine oil at the correct intervals. Other additive requirements for both diesel and petrol engine oils remain much the same.

What the Codes Mean

The oil's performance capabilities are specified by an API or CCMC coding somewhere on the package. The relevant codings for the former are: 'SE', 'SF' and 'SG' for petrol engines, and 'CC', 'CD' and 'CE' for diesel engines. The higher the second letter of each code, the higher the performance of the oil. Most modern diesel cars, including turbocharged models, require at least the API CD grade, while 'CE' copes better with stringent operating conditions.

CCMC specification codes are 'G4' and 'G5' for petrol lubricants, and 'D4', 'D5' and 'PD-2' and 'PD-3'* (PD standing for 'Passenger Diesel' - a specific car oil) for diesel oils. These codes cannot be directly translated into API gradings.

It may not be too relevant to quote the exact performance specification for each code here; suffice to say that your car handbook will provide the oil performance code required, as well as the viscosity range, and it's simply up to you to find the oil to suit these.

There is one other choice to be made, however, and that is between mineral and synthetic oils, or even the semi-synthetic type of oil that slips in between the two. Many people assume that synthetic oil necessarily represents the pinnacle of engine protection and that mineral oil is the poor relation. To a certain extent this is true, since synthetic oils - the products of chemical companies rather than oil companies, at dedicated chemical plants rather than refineries - offer certain obvious advantages over mineral oils. These include a longer life due to better oxidation stability (leading to a cleaner engine), a higher viscosity index (i.e. more stable viscosity over varying temperatures), better thermal conductivity (improved heat dispersal results in more efficient oil cooling), and better residual lubricity due to the longer adherence of oil to engine surfaces, providing more protection when the engine is started. Decreased volatility relative to the mineral stuff also implies that oil consumption may be reduced.

So, generally speaking, synthetic oils are better, but there are various qualities of synthetic oil available, just as there are on the mineral oil front too, and some of the better mineral oils subjected to high levels of refining, such as 'hydro-cracking', exhibit very high resistance to high temperatures and to oxidation. The cost of an oil is generally a good indication of its quality.

Saving on Synthetics

For many car drivers though, the benefits of a good synthetic oil may not justify the necessarily high purchase price, for those benefits are at a prime under rigorous operating conditions and in circumstances where maintenance is not carried out as often as it should ideally be. One great advantage to synthetic oil is its resistance to 'coking' (more easily comprehended if referred to as 'cooking'!) which is a boon in a turbocharged engine where the turbocharger oilways reach such high temperatures that they 'cook' mineral oil to form varnish and carbon deposits. This can appreciably reduce the life of turbocharger bearings, especially where oilways are obstructed by carbon build-up. The problem is not as great on a turbodiesel as on a turbo petrol, since the diesel's exhaust runs considerably cooler.

A handy halfway measure is presented by the semi-synthetic oil, a compromised blend of mineral and synthetic representing a mid- point in qualities. Its reason for being is that it is little more expensive than a good mineral oil, yet offers some of the advantages of synthetic - reduced volatility and improved low- temperature performance being among them. Semi-synthetics do have a real role to play and are often the best compromise between cost and performance.

The Fuel Injection Pump

The functions of the petrol engine's carburettor and ignition distributor can loosely be regarded as being replaced by the fuel injection pump. Why? Because it accurately measures the amount of fuel going to the engine (as does a carburettor), it determines the fuel injection timing (just as the ignition distributor sets the spark timing), and it distributes fuel to each of the engine cylinders (as the distributor cap and HT leads distribute volts to the spark plugs!). It is a mechanical engine-driven pump, which takes its drive from the same chain, gears

or (most commonly) belt that drives the engine camshaft.

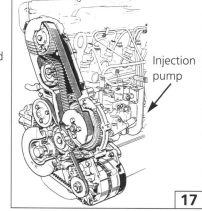

17. The Ford Fiesta and Escort 1.6 diesel's injection pump is driven by gears from the crankshaft, which in turn drives the camshaft via a conventional toothed belt. (Illustration, courtesy Ford Europe Ltd)

18. The Maestro/Montego MDi engine is one of many car diesel engines to have its injection pump driven from a conventional camshaft drive belt. (Illustration, courtesy Rover Group Ltd) As there is no throttle valve in the diesel's air intake, the accelerator pedal is connected straight to the injection pump, so power is determined by varying the amount of fuel delivered by the pump.

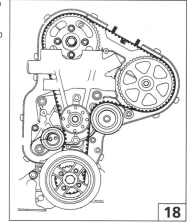

19. On most modern diesel cars, the injection pump also serves as the fuel supply pump, drawing fuel all the way from the fuel tank. Because this is a suction system, it is important not to allow any air to be drawn in via loose connections in the fuel lines, as air will impede injection, and in sufficient quantities, can bring the engine to a complete halt.

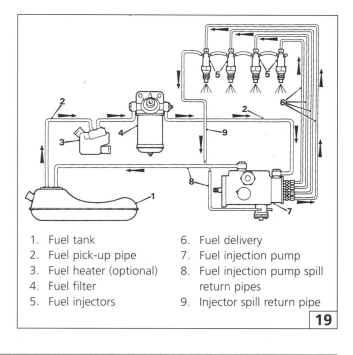

1. Fuel tank	6. Fuel delivery
2. Fuel pick-up pipe	7. Fuel injection pump
3. Fuel heater (optional)	8. Fuel injection pump spill return pipes
4. Fuel filter	
5. Fuel injectors	9. Injector spill return pipe

There is no manually operated cold-start enrichment mechanism on the diesel to equate to the petrol car's choke. Instead the injection pump automatically injects excess fuel under cold-start conditions. In short, there is very little for either the driver or the keen DIYer to get involved with! If you find a device on a diesel's facia that looks like a choke control, this is in fact a cold-starting aid which advances injection timing in order to keep cold-start smoke down to a minimum.

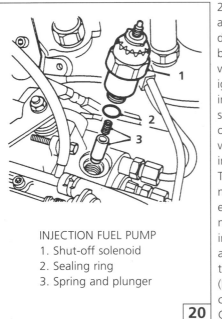

INJECTION FUEL PUMP
1. Shut-off solenoid
2. Sealing ring
3. Spring and plunger

20. It is often asked how a diesel engine can be switched off without an ignition system to interrupt. That's simple: you cut-off the fuel supply within the injection pump. This is done by means of an electric solenoid mounted on the injection pump and controlled by the ignition key. (Illustration, courtesy Rover Group Ltd)

The more curious and technically minded may wish to read the following brief description of the goings-on within a distributor-type injection pump, of which by far the two most common are the Bosch VE and Lucas DPC. We stress that this knowledge is not needed for DIY maintenance, and that you are under no obligation to learn pump workings by rote!

How The Injection Pump Works

21. The vast majority of diesel cars are equipped with a distributor-type injection pump...

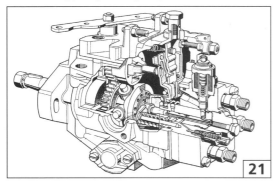

22. ...the minority are fitted with the in-line type common to heavy commercial vehicles. The fundamental difference between the two is that the distributor pump has only one high-pressure pumping element to serve all the engine's cylinders, while the in-line pump has a separate high-pressure pumping element dedicated to each cylinder. (Illustration, courtesy V.A.G.)

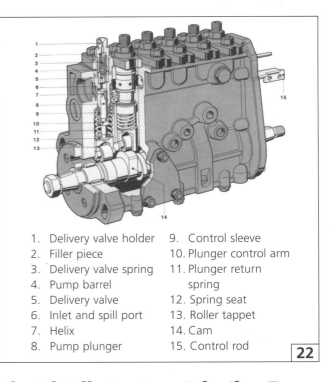

1. Delivery valve holder
2. Filler piece
3. Delivery valve spring
4. Pump barrel
5. Delivery valve
6. Inlet and spill port
7. Helix
8. Pump plunger
9. Control sleeve
10. Plunger control arm
11. Plunger return spring
12. Spring seat
13. Roller tappet
14. Cam
15. Control rod

The Distributor-type Injection Pump

The distributor pump's single element is operated by a cam ring or cam plate which has one cam per engine cylinder. Of course it also has one injection outlet per cylinder.

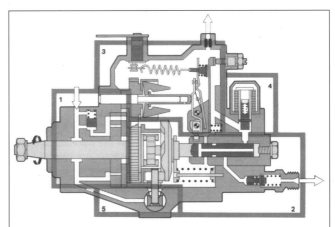

1. Vane-type supply pump. Supplies fuel from tank to injection-pump cavity.
2. High-pressure pump with distributor. Produces injection pressure, moves and distributes fuel to cylinders.
3. Mechanical governor. Controls engine speed, varies fuel delivery over control range.
4. Electromagnetic shutoff valve. Interrupts fuel delivery to stop engine.
5. Injection-timing unit. Adjusts beginning of injection according to engine speed.

23. All distributor-type injection pumps feature the following internal functions: vacuum-type fuel supply pump, injection-pressure pumping with fuel distribution to injectors, mechanical governing, injection timing variation, and engine shut-off.

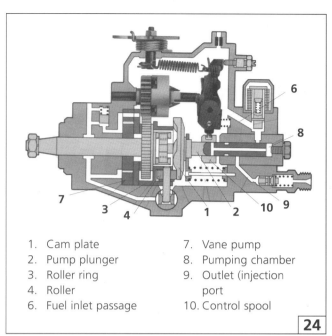

1. Cam plate
2. Pump plunger
3. Roller ring
4. Roller
6. Fuel inlet passage
7. Vane pump
8. Pumping chamber
9. Outlet (injection port
10. Control spool

24. In the cutaway of the popular 4-cylinder Bosch injection pump shown, the high-pressure pump/distributor head assembly converts the rotation of the pump shaft and cam plate (1) into fore-and-aft motion of the plunger (2). The roller ring (3) remains stationary while the cam plate and pump plunger are driven. The cam plate turns against the four rollers (4) in the ring, so forcing the plunger to slide rapidly fore and aft, pumping fuel to injection pressure (as high as 700 bar).

Fuel is supplied by the low-pressure vane pump (7) to the bore of the hollow plunger via an inlet passage (6). With the inlet passage uncovered, the chamber (8) charges with fuel. As the open end of the plunger moves back into this chamber, the fuel is pressurised and enters the plunger. As the plunger reaches the limit of its travel, fuel pumps at high pressure from an outlet port (9) to an injector, as an outlet to the relevant port becomes uncovered by a slit in the plunger.

At the opposite end of the pump plunger is a control spool (10), which is a sliding fit over the plunger. As the piston reaches the end of its travel, the spool (10) uncovers a passage in the plunger, letting the fuel out and relieving injection pressure. The position of the control spool is set by the driver via the accelerator linkage, but also by the action of a device known as the governor.

High-pressure outlet passages to which injector pipes are connected are positioned radially around the pump plunger. As the plunger turns, fuel is distributed at high-pressure to each one - in a similar manner to a rotor arm distributing a voltage to each segment of the distributor cap.

Injection Timing Mechanism

The injection purnp incorporates an automatic timing advance device (5) which compensates for the relative delay in injection and ignition as the engine speed increases. This device is controlled by fuel pressure in the injection pump, which rises

in direct relation to engine speed, turning the pump's roller ring slightly.

Injection Pump Shut-off

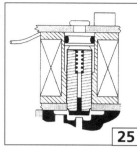

25. The only way a diesel engine can be switched off is by cutting its supply of fuel. With the distributor-type injection pump this is done by a solenoid plunger. The plunger is forced by a spring to obstruct a fuel supply passage in the pump. But when the car's ignition key is turned to `on', the solenoid is actuated by electric current, so pulling the plunger out of the fuel passage, allowing fuel to enter the high-pressure pumping section. (Illustration, courtesy Peugeot)

Injection Pump Governor

Engine speed is controlled partly by the driver (via the accelerator pedal) and partly by the governor (23.3). A governor is needed because the diesel engine is not self-regulating. This means that without a governor, if the load on the engine is increased, the amount of fuel delivered to it can go way beyond the amount which it can burn. The result is foul, black smoke at the tailpipe.

For a given engine load a maximum fuel stop can be fitted to the pump to prevent excessive injection, but if the load is removed from the engine (e.g. by going downhill) the simple fuel stop doesn't prevent engine speed from increasing unchecked, so the engine can literally rev itself to bits! Without a governor, the faster the engine runs, the more fuel the pump delivers and the more the engine accelerates... to the point of self-destruction.

And at the opposite end of the scale, towards idle speed, without a governor fuel output from the pump decreases as engine speed lowers, so the engine will just stop instead of idling.

The governor is a mechanism which automatically controls these two extremes of fuel delivery. It has centrifugal weights which are sensitive to engine speed, and by means of these it varies the quantity of fuel injected, so regulating engine speed. All engine speeds in between idle and maximum, however, are controlled directly by the driver's right foot.

The Injectors

26. The injection pump is worthless without the injectors, as these are needed to spray fuel into the engine cylinders.

Running from the injection pump are steel fuel pipes, one per cylinder, which pass metered fuel at high pressure from the injection pump delivery valve (a sprung outlet valve)

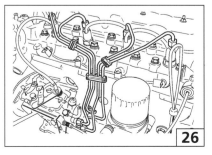

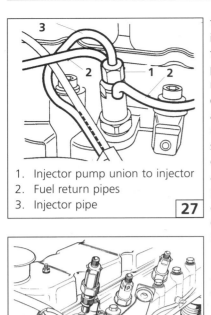

1. Injector pump union to injector
2. Fuel return pipes
3. Injector pipe

27

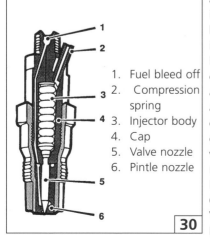

28

2

1. High pressure pipe union
2. Banjo bolt - leak-off pipe
3. Clamp bolt assembly
4. Pedestal
5. Injector
6. Seat washer
7. Locating ring

29

1. Fuel bleed off
2. Compression spring
3. Injector body
4. Cap
5. Valve nozzle
6. Pintle nozzle

30

to the corresponding injector. The three, four, five or six high-pressure pipes (as many as there are engine cylinders) have a high-pressure screw union at each end, similar to those found on brake pipes. This is a typical injector and pipe layout for a four cylinder diesel engine. (Illustration, courtesy Ford Europe Ltd)

27. The upper union connects to the injector holder. (Illustration, courtesy Rover Group Ltd)

28. This is either screwed or clamped into its bore in the cylinder head with a gas-tight seal between injector and head and this typical indirect injection installation also includes a heat shield. (Illustration, courtesy Ford Europe Ltd)

29. This is an alternative layout of the same thing! (Illustration, courtesy Rover Group Ltd)

30. Fuel enters the injector at the inlet (1) and passes down an internal drilling in the body (4) to a chamber in the nozzle (6). The holder combination comprises the holder and injector nozzle (6), and contains a vertical coil spring which acts on a pressure spindle which incorporates a nozzle needle at its lower end (5). (Illustration, courtesy V L Churchill/LDV Limited)

The injectors are linked by low-pressure steel piping or soft hosing (2, Fig.27 and 2, Fig 30) for the return of unused fuel to a point in the fuel system ahead of the injection pump (i.e. the fuel filter or fuel tank).

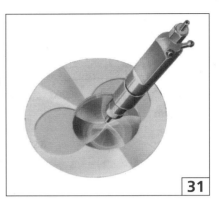

31

31. The injector holder secures the injector nozzle in the cylinder head and seals the combustion chamber. The nozzle sprays fuel into the combustion chamber under the control of fuel pressure resulting from the working stroke of the injection pump. Fuel entering the top of the injector under high pressure passes down to a chamber around the conical face of the nozzle needle. The fuel pressure in this chamber acts on the cone and forces the needle/spindle upwards against spring pressure. This causes the nozzle to open and fuel to be injected into the engine's combustion chamber. As injection comes to a close and pressure falls off, the spring returns the needle to its seat and injection is completed. A certain amount of fuel forces its way past the spindle and up into the spring chamber, from which it exits at low pressure into the fuel return line. This back-leakage of fuel is vital as it provides lubrication of the spindle, which can reciprocate in the nozzle bore many thousands of times per minute. (Illustration, courtesy V.A.G.)

That outlines the basic operation of the injector, but it's worth adding that there are different types of injector with varying characteristics determined by the design of the combustion chamber with which they are used. The formation of the fuel injection spray is greatly affected by the length, shape and size of the injector nozzle. If the opening is short, fine atomisation of the fuel is obtained, but with a long opening the spray has more penetration; a compromise invariably has to be made between these two situations.

32. In general though, there are two basic injector types: the orifice-type for direct-injection engines, and the throttling pintle type for indirect-injection engines. Common types of injector include the single-hole (32A), multi-hole (32B), long- stem, pintle (32C), delay and Pintaux types (32D), each of which is suited to a different design of combustion chamber. The last three are intended for indirect-injection applications. (Illustration, courtesy V L Churchill/LDV Limited)

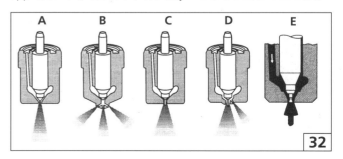

A	B	C	D	E

32

32E. With the pintle type a small cone extension on the end of the needle produces a pre-injection. As it opens, the nozzle needle initially permits the passage of only a little fuel through a very narrow annular gap (throttling effect). As it opens further, due to the rise in fuel pressure, the cross-section of the flow increases, and the main proportion of the fuel is not injected until towards the end of the needle lift. Combustion and engine operation become smoother with the throttling-pintle nozzle, because cylinder pressure increases more slowly. The pintle nozzle is the common type fitted to indirect-injection cars. (Illustration, courtesy V.A.G.)

With the multi-hole nozzles used on many DI engines, two or more small orifices drilled at various angles to suit the combustion chamber produce a highly atomised spray form. Many engines with direct injection systems use a four-hole nozzle with a high operating pressure of about 175 bar.

Two-Stage Injection Revisited

In the 1940s considerable attention was paid to pilot injection - the delivery of a small quantity of fuel in advance of the main charge so that the main charge enters a zone in which combustion is already initiated. The object of pilot ignition was to combat the phenomenon of diesel knock - too rapid a burning of accumulated fuel in the cylinder. Two-stage injection was experimented with, whereby the pilot spray was at a low pressure and the main spray at a higher pressure. Two-stage injection is at last being used to good effect on direct-injection engines to tame their production of noise and smoke at low speeds In this country, CAV Ltd. (now Lucas) applied the pilot injection technique in the form of the CAV-Ricardo injector - a unit designed specifically for use with the Ricardo Comet combustion chamber (detailed shortly). With this type, engine-starting conditions produce a small needle lift, so fuel passes through a small auxiliary hole in the nozzle and is directed to the hottest part of the pre-combustion chamber. At normal operating pressures, however, the full lift of the needle discharges the fuel through the main orifice. This nozzle design found great popularity due to its enhancement of cold-starting and low-speed running smoothness, and became known as the Pintaux.

DI and IDI - OK?

In case you remain unsure of the differences between DI and IDI engines, allow us to explain further.

Both systems of injection are used on diesel engines in current manufacture, IDI being almost universally employed for passenger car diesels, and DI for commercial vehicles. The fundamental differences between the two designs lie in the construction of the combustion chambers. The direct type has an injector positioned in the main (and only) combustion chamber over the piston, whereas the indirect type injects into a separate chamber in the cylinder head, known as the pre-combustion chamber. This communicates with the main chamber via a narrow passage. The main combustion chamber is usually formed in the piston crown, and the underside of the cylinder head is flat. Research work on the diesel engine in the 1930s led to a new and greater understanding of the complexities of the combustion process, and by the end of that decade these two fundamental

types of combustion chamber were emerging from the proliferation of available designs. British engineers concentrated their efforts on improving specific fuel consumption by increasing thermal efficiency through the use of direct-injection combustion chambers. Continental designers, however, were more concerned with achieving the smooth and progressive burning of fuel by a system of pre-combustion using an ante-chamber to originate a controlled flame which then spread to the main combustion chamber.

Indirect-injection

In the ante-chamber type of cylinder head, when the piston rises to the limits of the head, air is compressed into a chamber connected to the head by one or more passages. This chamber is maintained at a high temperature and a certain degree of air turbulence is created by the entry of air into it. The fuel injector is positioned in the ante-chamber and ignition of part of the fuel spray creates rapid expansion which forces the major part of the still unburned fuel particles through the hole or holes in the ante-chamber. The resulting finely divided spray of fuel in the main chamber then continues burning.

The combustion process in the ante-chamber engine was designed to give a more sustained push to the piston than would be the case if the burning of the charge took place in the open cylinder, for it was assumed that ignition in the diesel engine was more rapid than in the petrol engine. This was only true to a point; in the diesel engine there is a time lag between the commencement of fuel injection and the initiation of combustion - a prime cause of diesel knock . This was not clearly understood in the pioneering days of combustion chamber design, so the pre-combustion chamber was an attempt to control the combustion process.

Stopping Smoking

After further years of work it became possible to obtain controlled and complete combustion in an open chamber by means of more homogeneous mixing of fuel and air through better controlled turbulence, more refined injection equipment and improved fuel quality. It was found that smoke emissions from the exhaust were invariably caused by even the smallest amount of unburned fuel droplets, and the perfect matching of fuel spray pattern to combustion chamber shape and turbulence characteristics became an essential requirement in the avoidance of this phenomenon.

33. The greatest advance in combustion chamber design was made by Sir Harry Ricardo with the Comet combustion chamber for IDI engines. The Comet is an ante-chamber design in which the essential principle is the projection of fuel across a rapidly rotating mass of air in the combustion chamber, in such a manner that the air has to find the fuel rather than vice-versa. About 50% of the air volume is transferred into the cell or swirl chamber (pre-combustion chamber)

33

during the compression stroke, and the balance, after deducting that which is in the piston-to-head clearance, is continued in two shallow pan-shaped depressions machined tangentially to each other in the piston crown.

The air enters the swirl chamber to provide the necessary rotational movement. After partial ignition of the fuel in the chamber, the rush into the depressions in the piston crown forms a pair of oppositely rotating vortices. These bring the air contained in them into contact with the burning mixture issuing from the swirl chamber. The Comet head provided greater flexibility and freedom from smoke than any previous ante-chamber designs and was adopted by many prominent British engine makers.

Direct-injection

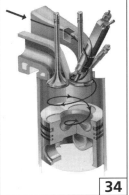

34. In an open-chamber (DI) engine, compression space can be reduced to give any desired compression ratio, the limiting factor being the valve-to-piston clearance. In the open-chamber design a high compression ratio is not necessarily required for ignition, nor even for cold starting, but in its simplest form the open-chamber does not give the required turbulence characteristics for the prevention of smoke emission. It does, however, have the advantage of diluting the fresh air charge less with residual exhaust gases.

Thus the DI engine is efficient from most aspects, it is a reliable cold starter, it returns good specific fuel consumption because of superior thermal efficiency (typically 20% better than the IDI engine) but it is nonetheless inclined towards roughness in operation and incomplete combustion. The significantly greater economy of the DI is one of the main reasons for its dominance in trucks, in which applications operational refinement is not of paramount importance.

Some of the most important considerations in the open-chamber engine are connected with fuel injection (another disadvantage of the DI is its reliance on considerably higher injection pressures than are required in the IDI), as the pressure of injection, the locations, penetration, shape and direction of the fuel spray are all critical to efficient operation.

IDI allows smoother, more progressive burning of fuel and so quieter, more refined and often cleaner engine running - hence its near-universal application to passenger cars. But it returns comparatively poor economy because the addition of the pre- combustion chamber to the main chamber substantially increases the surface area of the total combustion zone, and that means that more useful combustion heat is dissipated to the cylinder head than is with the DI design. Operational efficiency is also reduced by the pumping losses incurred by the air charge passing in and out of the pre-combustion chamber via a narrow passage. With recent advances in combustion chamber and injection system design however, the practical differences between DI and IDI are narrowing, particularly with developments in

fuel-injection technology, turbocharging and intercooling.

The Fuel Filter

35. The internal workings of the injection pump and injectors are machined to give such minuscule operating tolerances that any impurities carried in the fuel will rapidly wreck these expensive components. An injection pump can be turned to scrap by just 5 grams of microscopic dirt!

Diesel fuel acts as both lubricant and coolant for the moving parts of the injection pump and injectors, so just as it's vital to keep engine oil filtered for the engine's sake, so it is essential to filter the fuel that passes through the injection system.

Every diesel car has a fuel filter in its engine compartment, connected in line between the fuel tank and injection pump. Usually it's mounted away from the engine, but sometimes it's actually on it. The filter comprises a pleated roll of special paper capable of trapping particles of 5 to 20 microns in size.

How Does It Work?

36. In a typical design of filter, fuel entering the filter head passes through a hole at the top of the filter element and travels down through the pleated paper into a bowl. From here it is forced up the centre tube of the element to the outlet in the filter head. As it passes through the paper it gives up contaminants to it. (Illustration, courtesy V.A.G.)

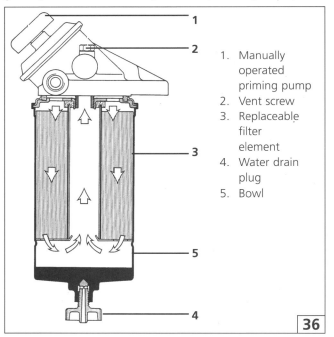

1. Manually operated priming pump
2. Vent screw
3. Replaceable filter element
4. Water drain plug
5. Bowl

37. The filter (4) eventually clogs up with dirt, so it has to be renewed at specified intervals before fuel-flow is reduced or unmanageable chunks of dirt end up being passed to the injection pump. (Illustration, courtesy Rover Group Ltd)

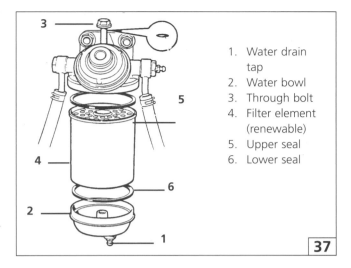

1. Water drain tap
2. Water bowl
3. Through bolt
4. Filter element (renewable)
5. Upper seal
6. Lower seal

37

Removing Air & Water

38. But this isn't all the fuel filter does. It also provides a means of bleeding air out of the fuel system, via a vent screw. This is necessary because any air present in the fuel lines causes problematic running. Where there is air in the system, there is

38

no fuel, and the system requires a continuous, unbroken flow of fuel to keep the engine running. If there is a large enough air-lock the engine might not even start at all.

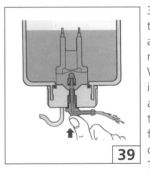

39

39. Water is similarly damaging to the injection pump and injectors, as it can corrode their finely machined internal components. Water can be found in diesel fuel, if only because it's impossible to avoid condensation in a car's fuel tank. So provision is made for the fuel filter to condense water out of the fuel and trap it in its bowl. This happens naturally, as water is heavier than diesel fuel, so it sinks to the lowest point in the filter assembly. This type has an easy 'push-to-open' water drain. (Illustration, courtesy V.A.G.)

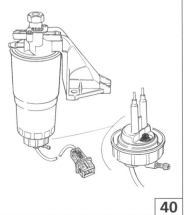

40

40. Water should be drained from the bowl at the specified service intervals. Many modern filters include a water-in-fuel sensor

which illuminates a warning light on the dashboard when draining is required. (Illustration, courtesy V.A.G.)

Heating The Fuel

Why would you need to heat diesel fuel? Because it has a high wax content which crystallizes at very low temperatures and can clog the fuel filter. The result is a misfiring, stalling or even non-starting engine. The application of a little heat to melt the wax formation soon gets things on the move again though.

41. Although modern winter-grade diesel fuel has additives to prevent waxing down to temperatures as low as -20 degrees C, many diesel cars are fitted with a fuel heater mounted either within or around the fuel filter or in the fuel supply line near the filter. This switches into and out of operation automatically by means of a thermostat.

41

42. An electric fuel heater can usually be recognised at a glance by the presence of electrical wiring to the heater unit (A), but coolant-type heat exchangers (usually bolted to the cylinder head) are less obvious.

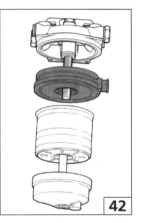

42

Volkswagen-Audi circulation-type fuel-heater

Rather than employ a separate, powered heater, the Volkswagen-Audi group prefer to keep the fuel warm by circulating it through the pump. The pump, naturally, is hot, since it is bolted to the engine. At the top of the VAG fuel filter - which doesn't have a filter head, as do more conventional filter assemblies - is a pre-heater valve. Heated fuel returning from the injection pump and fuel injectors flows back to the fuel filter via this valve.

Depending on the ambient temperature the pre-heater valve either directs the warm fuel back to the fuel tank or to the suction side of the fuel filter to prevent the fuel from becoming viscous from wax formation.

At ambient temperatures above 10 deg. C the control valve is at rest and the fuel flows back to the tank without pre-heating the filter. At ambient temperatures below 0 deg. C the control valve closes the return line to the fuel tank. The heated fuel opens the spring-loaded non-return valve and flows straight back to the fuel filter suction side.

The Pre-heating System

Just as Ackroyd Stuart's engine needed some initial heat to get it running, so the modern diesel benefits from extra heat when firing up from cold. Although the compressed intake air gets very hot right from the first stroke, the cold cylinder walls soak that heat away, so the addition of a "glow plug" in the pre-combustion chamber helps compensate. Direct Ignition engines don't usually need a pre-heating system. However, some DIs do have glow plugs.

43. The glow plug is a long, thin electrical heating element which protrudes into the pre-combustion chamber.

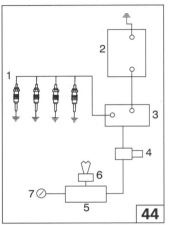

44. It is connected electrically to earth via the cylinder head, and to a positive voltage supply.

When the driver turns the ignition key to the "pre-heat" position, the glow plugs heat up to about 800- 900 degrees C. After a few seconds of pre-heat the engine can be started.

An electronic pre-heat control unit ensures that the glow plugs are on for no longer than is necessary, and a pre-heat warning light at the dashboard extinguishes to inform the driver that the engine can be started. The whole process takes anything from 2 to 20 seconds depending on engine type and model.

Turbocharging

45. Diesel engines are turbocharged not only to increase their performance, but also to improve smoothness and decrease smoke emissions. By turbocharging an engine you pack more air into the cylinders, which allows more fuel to be burned and so more power to be produced. (Illustration, courtesy Ford Europe Ltd)

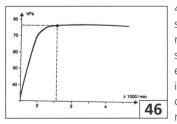

46. Traditionally, turbochargers suffer from "turbo lag" - reluctance of the turbo to spin and producing an engine with very little 'go' in it from low speeds, until a certain point in the rev. range - at which point the older, cruder turbocharged petrol-engined cars would take off alarmingly! But turbo lag is less of a problem in diesels than in petrol engines, since the large, unthrottled volume of gas which passes through the engine at all speeds keeps the turbocharger spinning even at low speeds. (Illustration, courtesy Ford Europe Ltd)

47. About 40 percent of the useful heat energy generated by the unturbocharged engine is wasted in the form of hot exhaust gases. But as the turbocharger is driven by some of this otherwise wasted energy, the operating efficiency of turbo diesels is higher than that of non-turbo engines. (Illustration, courtesy Ford Europe Ltd)

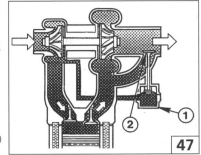

Intercooling

48. Intercooling is often applied to turbodiesels to decrease intake-air temperature. As the act of compressing air (which is what a turbocharger does) makes it heat up and expand, this to a certain extent coun-

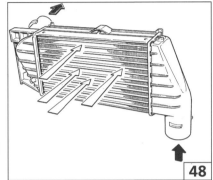

teracts the benefit of increasing its pressure, so it's useful to 'inter-cool' it - cool it downbetween turbocharger and cylinders. The intercooler is located in a cool air stream, usually at the front of the vehicle, so it gives up heat from the intake air to the atmosphere. Without an intercooler a turbocharger can still be of great benefit. With one, however, it can give even more boost, so producing further power. (Illustration, courtesy V.A.G.)

Brake Vacuum Pump

A vacuum pump is needed because the absence of a throttle valve in the diesel's air intake means there is no vacuum to tap into for the servo. Brake vacuum pumps need no maintenance - other than occasionally to the drive belt, if of the belt-driven type.

49A. This type is actuated directly by a cam on the camshaft.

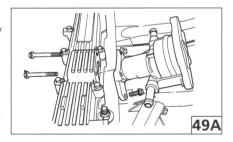

49B. This one is belt-driven from a pulley. Many are driven directly off the end of the camshaft.

CHAPTER 4 - BUYING A DIESEL CAR

"Diesel engines last for ever. So I don't have to worry about the condition of the car I'm buying, do I?"

Well, actually, it's 'no', and 'yes', in that order! While diesel cars certainly do generally last longer in the engine department than their petrol-powered brethren, they also tend to have more hard use. And what's more, not all diesel engines are equal, and some will appeal more to certain owners than to others. We set out the options and the areas to check when purchasing diesel power.

Part 1: CHOICES

Which Car Do You Choose?

Before taking the plunge, and buying a diesel car, you should decide exactly what you want from the car. Important questions to ask yourself are: how many people are likely to travel in it? Will journeys be mostly long or short? Do I need a hatchback or an estate? Do I need a turbodiesel?

If you may be doing a lot of motorway driving you should ask yourself: Is the car relaxed at motorway speeds? Has it got five gears? Would cruise control be an asset? Which is more important: power or economy? Would I benefit from an automatic transmission? (Rare with a diesel, but not out of the question.)

How important to you is power steering? Bear in mind that diesel engines are invariably heavier than the petrol engine fitted to an equivalent size of car, making even small cars heavy to steer at parking speed. If you're reading this book then it's fairly safe to assume that you're a keen DIYer intending to do your own servicing work, and maybe repairs too; if so, you may wish to keep clear of complicated vehicles such as Citroen's CX or the Peugeot 505!

Those are the everyday practicalities of the car, but there are less obvious considerations to be made. For instance, do you intend to keep the car for many years? If so, you can better justify opting for one of the bigger depreciators offering more car for the money, since you won't have to put up with a major financial loss a couple of years down the road, and you'll be bagging yourself a better deal. As a general rule of thumb, the larger and more lavish the car, the greater the depreciation, and that rule applies whether you're talking petrol or diesel.

Insurance & Fuel Consumption

Don't wait until you have bought the car before finding out how much it will cost to insure. You shouldn't assume that because it's a diesel, it will be both low in performance and unattractive to car thieves, and therefore cheap to insure. These are common misconceptions, coupled with the fact that many insurance companies still labour under the delusion that diesels are more expensive to repair.

Do make a point of considering fuel consumption before-hand. Whilst it is true that fuel economy is one of the major attractions, not all diesels are frugal.

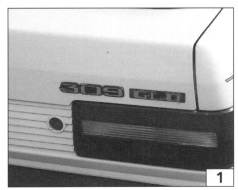

1. You can identify the model-spec. by the markings on the back of the car, in most cases - but see below for how to find out which is which.

Where Do You Get Facts & Figures?

2. Having sorted out the ideal car type for you, you need to find out which comes equipped with a diesel engine,

and which of those are worth having. Probably the best source of information is the Test Data section of Diesel Car magazine (FIG.2). In it you'll find facts and figures on all the diesels tested by them since 1988, along with the number of the issue each road test appeared in. Back copies of road tests are available.

If you're intending to buy a new car, you'll want to refer to the Data pages in the latest issue of Diesel Car. This is the only specialist magazine currently available for diesel cars. In order to compare figures with the petrol-engine equivalents, look up the road test results in Autocar or one of the buyers-guide magazines - see below.

Finding a Diesel

Buying a second-hand diesel requires more application and investment of time (and, unfortunately, money) than buying a petrol car. You can't rush the process, and as well as this being a frustration, it can cost several pounds a week just to buy the journals that carry the right car classifieds.

So where do you start looking? It can be quite difficult to find second-hand diesels, for a number of reasons. The first is that it isn't obvious where to start looking for one: while local papers are stocked with petrol-powered cars, there are usually relatively few diesels. The second difficulty is that diesels are often uncategorised, and lost among the plethora of "conventional" petrol cars.

3. You should take a look at your regional issue of Auto Trader, and Motoring Exchange and Mart. The former contains classifications by vehicle type (e.g. Saloons, Estates, Hatchbacks and - usefully - Diesels) while the latter lists cars by make, so they're both more efficient to use than local papers, although they both cover a far larger geographical area.

The back pages of Diesel Car magazine are filled with advertisements from dealers who specialise in second-hand diesels, as well as private classifieds. You will also find dealers advertising in local papers and listed in Yellow Pages and in similar directories, so if you're buying new and you know that you're interested in, say, a Peugeot diesel, there's absolutely no problem finding where to go.

If you do manage to find a suitable car advertised, the chances are it will have been sold by the time you make your enquiry, so popular and comparatively scarce are good used diesels. In any event, to avoid too much disappointment, expect to lose the first few cars you find. You should be prepared for the whole process to take quite a while.

New or Used?

Most cars lose between 25 and 33% of their value in the first year alone, yet they are still virtually new. You can expect one-year old cars to be in excellent condition, free of corrosion

and (particularly with diesels) hardly run-in: the mileage on a one-year old's odometer will typically be 12,000-25,000 miles.

Buyers should bear in mind the fact that a one-year old car which has already depreciated handsomely will continue to do so, although the scale of loss will be smaller. And that the scale of loss is invariably smaller with a diesel than a petrol car - which is another reason why buying diesel makes sense.

A brand new car gives you the security of a warranty valid for anything from 12 to 36 months, depending on make, and there are often other carrots such as free breakdown recovery. Such cover can be separately bought for a second-hand car for a lot less than the depreciation losses on a new car. (In other words, you don't need a new car to benefit from such breakdown cover.) You will need a car with a totally complete service record and a low-ish mileage. Visit your local main dealers and diesel specialists, find out what they charge for an extended warranty (you don't have to have bought the car off them - they'll make a profit out of selling you the warranty!) and you'll know what sort of car you'll need to buy in order to be able to buy yourself a warranty to go with it.

Price Guides

4. If you've decided to buy used, you will need to find out the value of the car of your choice. The motor trade refers to Glass's Guide and the CAP "black book" for guideline buying-in and selling-out prices (FIG.4). These books are not made available outside the trade, but there is an abundance of independent car price guides for sale at newsagents' shops.

5. None of the price guides available through newsagents is gospel: compare prices between guides and you will find variations - some quite large - and few of them take into account regional variations in value. But if you refer to two or more different guides you can certainly gain a worthwhile feel for a car's second-hand value, especially if you compare theoretical guide prices with the "real" ones quoted in advertisements.

Something else pocket-sized price guides are good for is providing a potted history of each car range - useful for documenting modifications, equipment levels and model-range breakdowns as well as details and dates of major improvements, such as ABS (anti-lock brakes) or air bags as standard fittings.

Glass's Guide contains a 12-point definition of what it considers to be "Guide Condition Retail", and it is these minimum requirements that should be met in order to achieve the "GCR" value (theoretically the value on the car's windscreen if it's a trade sale). In a nutshell, what it says is that the car should be free from rust and dents, that every device or system on board should be functioning correctly, and that certain items, such as exhaust, clutch, brakes, tyres etc. should not be in imminent need of replacement. If anything is malfunctioning or on its way out, that is reason for negotiating a reduction in price.

Who Do I Buy From?

6. If you have chosen to buy new you have little choice but to use a franchised dealer, although you can choose whether to buy or lease.

7. Ford's Options plan, for instance provides alternative ways of 'owning' a new car - but you do need to work out all the cost implications for yourself, and remember that a guaranteed trade-in figure will only be available if you stick with the same manufacturer.

If, however, you are buying second-hand, there is a much wider choice. The obvious one is to buy a used car from a private seller, although that's a fairly risky business offering virtually no recourse if you make a mistake; the onus is on the buyer to establish whether or not the car is good before he/she purchases it. If you are confident when it comes to assessing the genuineness of the car, that's all well and good, otherwise you should tread with caution.

If the vehicle is obviously misrepresented by the private vendor, you do have some recourse if you can prove the misrepresentation, so it's always a good idea to get it in writing that the vehicle is as described. Also, if it proves unroadworthy, that's a criminal offence, and the long arm of the Law might be persuaded to reach out and prosecute someone ... eventually!

Do remember, however, that you can always try to make your private purchase subject to inspection by a specialist. The AA and RAC will carry out an inspection for a (hefty) fee. A main dealer near to the vendor might be a better bet: give them a 'call first - they might be quicker, cheaper and give you a better opportunity to give the car an instant Yea or Nay.

8. Buy secondhand from a trader, however, whether a large dealership or a sole trader, and you have legal protection in the form of the Sale of Goods and Trade Descriptions Acts,

which state that the goods (in this case the car you are buying) must be fit for the purpose for which they are intended, must be of merchantable quality, and must be as described. Following through a prosecution under these acts can be very difficult though, and if the prosecution is successful, this is no guarantee that you'll get your money back. In practice, legal protection is worth little, except as a last, desperate resort, so once again, ensure that you have an inspection carried out by an independent specialist.

There is also a legal obligation for a trader to put right anything serious that goes wrong with a car within a certain period of time after he has sold it to you. Unfortunately that period of time is not specified and remains open to debate.

If you buy from a dealer of substance rather than a back-street type you will find the terms and conditions of sale generally include some specified warranty period provided automatically, although you are usually paying for it in the form of a paid-for warranty, the cost of which is included in the purchase price. Any repairs covered by the dealer are at the discretion of the dealer, so it's always wise to enquire before committing money. Also, insist on taking away and reading carefully the warranty supplied. There are usually a lot of components not covered by it!

Generally speaking, the larger the dealer, the less likely you are to be hoodwinked, but the more money you are likely to be parted from. Small dealers with fewer overheads, who throw less "goodies" into the package, often offer advantageous prices. *(Publisher's Note: It is not unknown to be hoodwinked by a very well known dealership, however, as we know from personal experience. Our advice is to trust motor traders about as far as you can throw their showroom until proven otherwise!)*

Choose a dealer who is a member of the Retail Motor Industry Federation (identified by the cogwheel motif) and you'll find that a warranty is automatically provided on any car with less than 60,000 miles on its clock. Do take note of what was said earlier about the restricted nature of warranties. Seals may be covered, but for parts only, not labour. So, stripping an engine to replace a seal can still cost you hundreds of pounds. Look out for 'clever' get-out clauses on the part of the insurance company which may have prepared the 'warranty'.

9. However, you might take heart in the fact that the used forecourts of many franchised dealers have some sort of protection built in for the buyer, other than the obvious legal protection already mentioned. That's

because used-approved car sales systems such as Rover's approved used car warranty programme, Vauxhall's Network Q, and Ford's Direct, ensure that a points-based inspection system is carried out on any car intended for the forecourt, and no car makes it to the showroom or forecourt until all points are satisfied.

Buying at Auction

10. Auctions are valid alternatives to dealers, individual traders and private sellers. For most private buyers, though, they are the riskiest place from which to buy, even though they can be the cheapest source of used cars.

The majority of auction-goers are professional traders, because it's from auctions that the trade sources much of its stock. To the private buyer, the auction game is all about putting in the bid when the traders won't bid any higher; they cease the bidding when their profit margin starts to erode. You don't get the car at trade price, but you should end up buying at a price that's well short of the forecourt price, and also less than the price of a private sale.

But auction buying is fraught with problems, the process of sale happens in something of a rush (sometimes in under one minute!) and it's all too easy to make a regrettable decision when the pressure is on. You don't have a chance to drive the car beforehand, and if you buy a car described as "as seen", you have no recourse if it proves to be a banger. Otherwise you usually have an hour in which to assess it, find major faults, and cancel the deal if need be. At an auction you may well purchase a car without MoT or road fund licence, whereas dealers will take care of such annoyances for you.

You can certainly bag bargains if you learn how to play the game, but you don't get any of the convenience of a dealer purchase. Never attempt to buy at auction without first having observed three or four of them in progress. The speed of proceedings and the gabbled chat of the auctioneers really do pose a threat to the novice auction-goer.

The good news is that both of the biggest auction houses, ADT and Central Motor Auctions, hold diesel-specific auctions about once a month or once a fortnight. The auctions are mostly stocked with diesels from six months to four years of age, turned out by company fleets and hire companies. There are often excellent cars among them.

For Your Protection

By now you have decided on the car you want and you have a good idea of the price you'll be paying for it, so what comes next?

One thing often overlooked when buying a new car is the matter of its authenticity. This is an oversight which is

catching out an increasing number of people, so it's something of which you should be well informed.

You need to be sure that the vehicle is free from outstanding finance or HP arrangements, and is not stolen. The biggest possible shock you can receive, having committed your money, is to discover that the car wasn't the seller's to sell in the first place. This actually occurs more often than you'd think, and can leave you with nothing to show for your expenditure, and no means of retrieving your financial loss. In other words, under the law as it operates, you lose both the car and your cash!

This is why it is advisable to ask the seller to declare in writing that the car is free from hire-purchase or lease-purchase arrangements. Also, if the car is being sold on someone else's behalf, that the seller is authorised by the true owner to sell the car - he should have a signed note of authorisation from the owner. None of which, of course, is by any means fool-proof. Even if the car was sold in good faith, if it is found that the car was stolen three 'owners' ago, you end up the loser.

There is quite a high risk of buying a used car which has been subjected to a major damage-related insurance claim. This could be positively dangerous to the buyer, as well as to his wallet. A crash-damaged-repaired car might have been dangerously repaired and will certainly be worth less, whether or not rebuilt around a replacement bodyshell. You can reduce this risk *(but not eliminate it! - Publisher)* by shopping only with a reputable motor dealer - one bearing the hallmark of the Retail Motor Industry Federation is a safe bet. You can only be sure by having an AA or RAC inspection carried out. Not only will they check that all 'numbers' are correct (see below), they will also be aware of all the tell-tale signs of a car that has been in a smash.

HPI Autodata - a Check on Authenticity

If you are buying from an entirely unknown quantity, you should exercise real caution by applying to HPI Autodata for a thorough check on the car's authenticity. For a moderate fee, HPI will check (either by post or over the phone) that there are no outstanding hire- or lease-purchase agreements, and also that the seller has clear title to the vehicle. HPI has been in service on behalf of the trade since 1938; now its service is available to everyone.

Investigations carried out for private buyers have revealed that 1 in 5 cars have been the subject of a damage-related insurance claim at some time, 1 in every 6 have an outstanding financial agreement recorded against them, and 1 in 140 cars have either been reported as stolen or are declared by HPI as being much at risk from theft. HPI Autodata can be contacted on 01722 412746.

With such valuable groundwork covered for you, you must now ensure that the chosen car is as physically wholesome as it initially appears. Unless you're mechanically minded or have a knowledgeable friend who is prepared to accompany you when you're looking for cars, you either take the risk of losing the car by taking the time to arrange a professional inspection, or you buy only from a reputable dealer. But if you're prepared to take the risk of losing the car, who are the people you can turn to in your hour of need?

Professional "Once Overs"

11. As we have mentioned earlier, both the AA and the RAC offer mobile vehicle inspection services and you don't have to be a member to take advantage, although you should be aware that inspections can cost from £100 to £300, depending on the type of car. Many garages also 'hire-out' mechanics for inspections.

The cost of a professional inspection of this sort should be considered in relation to the amount you're spending on the car. Also worth thinking about is what to do if the first couple of cars you have inspected fail their inspection - how many such checks do you pay for? Fortunately, AA engineers apparently advise people not to buy in only 12% of cases, although in many instances they recommend purchase only if certain essential repairs are first carried out.

If you don't want to pay dearly for an inspection, you can find out which dealers in your region contract AA or RAC engineers to carry out random spot-checks on their used cars. This is a system of self-regulation which maintains a high standard of second-hand product.

Failing all else, it is worth telling the salesman you will buy the car you have selected providing he pays for an AA or RAC inspection on the car, and carries out any specified repairs. It's a relatively inexpensive way for a salesman to guarantee a sale, so the suggestion will probably be taken up.

Part II: ASSESSING A DIESEL ENGINE

Introduction

As diesel cars are no different from petrol-engined cars in all areas other than the engine and fuel system, they can be checked over in the usual manner. In this section we're looking at only those aspects which differ from the petrol cars, with which most people are familiar.

Some of the following suggestions may look unrealistic, in that you can't strip down components on a dealer forecourt or in a stranger's driveway. That's agreed, but this information could be very useful if ever you're buying from a friend or relative who doesn't mind you "poking around".

12. The diesel engine has a justified reputation for durability,

and that's what keeps the residual values of diesels so high, including the Citroen ZX Turbo Diesel shown here. One reason they last so well is that they are more robustly constructed, because they have to withstand greater cylinder pressures. Another is that diesel fuel is a lubricant, unlike petrol, and is consequently less wearing on cylinder bores and pistons. A third is that, because diesel engines don't rev as high as petrol engines, their rate of wear is further decreased. The upshot is that, as a rule of thumb, a well maintained 150,000-mile diesel usually shows less wear than a well maintained 100,000-mile petrol engine. Symptoms of wear, such as piston slap and oil burning, are far less common, and some diesels reach the 250,000-mile mark without so much as a cylinder head overhaul!

Moreover, when you take away the ignition system, you get greater inherent reliability. There is also less to maintain and "tune-up" on diesels, although they do require more frequent engine oil changes. You are less likely to find faults with diesel engines than with petrol engines, and where you do they are often of a diesel-specific nature, related to the fuel system.

Some car owners and buyers are reasonably confident about listening to and trying out second-hand cars with a view to a purchase. You don't have to have been around old bangers for long to know what sounds good and what doesn't. Diesels tend to be a different matter though - most notably because they make the wrong sort of noises to start off with! As we've already said, the differences aren't really that great, although there are some specific things to watch out for.

13. As most fundamental engine components are the same as in the petrol engine, problems of worn main bearings, knocking camshafts or big-ends, rattling small-ends or timing chains certainly still occur as a result of very high-mileage use or skimped servicing. But the busy clattering sound of diesels can go some way towards concealing these tell-tale noises.

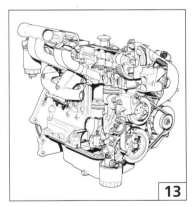

Most faults are likely to be connected with the operation of the fuel system, as this directly determines engine operation, although malfunctions of the engine pre-heating system aren't uncommon, and they can make an engine difficult, noisy or smoky to start from cold.

We suggest you use the following advice in tandem with the fault-finding information given in *Chapter 6*.

Engine Oil

As the diesel has fairly stringent requirements when it comes to engine oil and filter changes, it's important to ascertain that it has been properly maintained. That means asking for a dealer-stamped service history book - one that is complete and not a later 'replacement' book. If you are given excuses in place of a service history, assume the worst and, if the car is fairly new, walk away especially with a turbo engined car.

14. You can realistically expect the oil to look black, unless it has very recently been changed. That's okay, but if it appears sludgy, it probably hasn't been changed in a long, long time.

15. If there is evidence of beige mayonnaise-like froth on the dipstick and under the oil filler cap, a cylinder head gasket failure is possibly indicated, so steer clear.

16. Oil droplets on the inside of the radiator cap are another indication of cylinder head gasket problems.

Cold-Starting

When assessing the state of a diesel engine, it is important to start it up from stone cold, since inadequate cylinder compression (due to internal wear) will make a cold start difficult.

Compared with the petrol engine, there are relatively few things which can prevent a diesel engine from being started from cold. The most likely causes are poor cylinder compression, a malfunctioning pre-heating system, a weak battery, malfunctioning injectors, and a lack of fuel. Poor compression can be due to piston/cylinder wear, or worn or incorrectly adjusted valves. The first of these is very expensive to rectify, while the repair of worn valves requires cylinder head removal, and so is also costly.

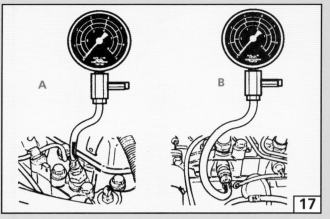

17. Ideally carry out a compression test (always) on a hot engine. Any cylinder showing a compression reading that is much more than 15% lower than the highest reading can be suspected of having serious wear. Figure A shows the tester connected to an injector location; B using the glow plug hole adaptor. IMPORTANT NOTE: See *Chapter 6, Fault Finding, Job 4A* for further information on testing cylinder compressions. (Illustration, courtesy V L Churchill/LDV Limited)

Pre-Heating System

The cold diesel engine requires cylinder pre-heating by means of glow plugs. There is one plug per combustion chamber, supplied with voltage from a control circuit when the car's ignition key is in the "pre-heat" position. A pre-heating warning light is illuminated at the instrument panel when the system is activated, and it stays on for a predetermined time, during which you shouldn't really attempt to start the engine. The duration of the pre-heating period varies from car to car, and can be anything from 2-20 seconds. Check the handbook of the car you are examining to find out what is considered normal for that make or model.

If the car does not start after the warning light extinguishes, put it into pre-heat mode a further once or twice. If it starts more readily it could be that one or more glow plugs are out of action. If it doesn't, it's likely that the pre-heating isn't effective.

18. You can quickly check for a dead pre-heating circuit by connecting a 12 volt test bulb between the centre terminal of the first glow plug (the one with the supply lead to it) and earth (the cylinder block).

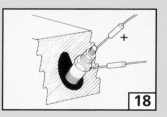

19. If it illuminates when the ignition key is in pre-heat position, you know at least that there is no problem with the supply.

It's often easy to remove the glow plugs for inspection, (although some are difficult to get at) by disconnecting their connecting lead or bar and unscrewing them from the cylinder head with a spanner. Pitted or eroded plugs must be renewed and may indicate a fault with the injector or the spray pattern.

Starting System

If the car won't start, a starter motor that is turning sluggishly could well be the problem. Unlike a petrol engine, a diesel will not fire from cold, without a good cranking speed. Connect a voltmeter across the battery terminals as the starter is operated; if the voltage reading drops much below 10 volts then either the battery is in a poor state or there is a problem with the starting circuit.

Black Exhaust Smoke

20. The first thing to check once the engine has fired, is the exhaust. From most diesels you can expect a substantial puff of black smoke upon first firing up, but if the black stuff continues to pump out, something is wrong. Black smoke is unburned fuel, and it's a possible indication of faulty injectors (often accompanied by loud knocking noises), a dirty air filter, incorrect injection timing, or an incorrectly adjusted or faulty injection pump. Details of how to check injectors are given in *Chapter 5* (along with more in-depth testing of the pre-heating system). Checking the cleanliness of the air filter element is plain and simple. The injection pump, however, is far from straightforward, and requires the intervention of a diesel specialist.

Black or dirty grey smoke tends to be more common on the direct-injection (DI) diesel engine than on the almost universal indirect-injection (IDI) type. It is usual to see more dark smoke from the former type of engine than from other (IDI) diesels. If you are not sure just how much smoke is acceptable, seek the opinion of an expert.

Blue Smoke

21. A diesel engine will produce blue exhaust smoke if it is burning its engine oil - as will a petrol engine. This is either a piston/cylinder wear problem, or wear of the valve stem oil seals/valve guides - cheaper to put right, but still costly and inconvenient. Quite a few diesels, especially of the DI type, emit some blue smoke while they are warming up. This is okay provided it stops when the engine reaches operating temperature.

Lumpy Cold Running

Look at the engine when it has just started from cold: if engine running sounds lumpy, and the engine is swinging energetically on its mountings (you'll probably find black smoke coming from the exhaust as well), beware. These symptoms testify to poor compression or poor injector performance on one or more cylinders. If the engine smooths out as it warms up, that is a further indication that there is a cylinder compression problem, so the car should be avoided.

Crankcase Ventilation

22. Another sign of piston and cylinder wear is crankcase over-pressurisation due to blow-by of gases from the combustion chambers. You will find a large hose venting the crankcase, running from somewhere on the sump or cylinder block, into the air filter casing or intake manifold. Disconnect the upper point of this hose while the engine is running at normal operating temperature, and see if there is a thick oil mist pumping from it. If there is, and there is oil build-up in the intake manifold and/or air filter casing, the engine is probably worn. A slight oily film is passable, but any more is undesirable.

Oil Leaks

23. Heavy oil consumption, as well as being caused by oil combustion, can be due to oil leaks from the engine. Look out in particular for oil leaks from behind the timing belt cover. If you spot any, tread with caution, as oil degrades cam belts, and a failed cam belt will wreck a diesel engine.

Exhaust Colour and Head Gasket

When a diesel engine is running okay at normal operating temperature, you can expect the exhaust emissions to be virtually colourless. However, if it is white or grey/white, this could be due either to steam (you can differentiate steam from smoke because it disperses more quickly) caused by coolant leaking into the cylinders (a damaged cylinder head gasket) or to "quenched" combustion. Double-check oily deposits floating on the coolant.

Quenched combustion means that the combustion process has started but has been aborted prematurely for one reason or another (too cool a combustion chamber, incorrect injector spray pattern or opening pressure, incorrect injection timing etc.) - so it really does warrant investigation. White smoke/steam up before the engine has reached normal operating temperature is not ominous, although it might be an indication that the glow plugs are not working as well as they should.

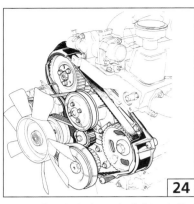

Camshaft Belt

24. You can't afford to ignore the age or condition of the camshaft belt on any diesel. If the car has covered 50,000 or more miles, check the service record for evidence of camshaft belt renewal (assuming the

24

camshaft is belt-driven). If this vitally important component has not been renewed at this mileage, and the vendor is a trader, press for it to be renewed as part of the deal, as failure of the cam belt will cause costly internal engine damage. (Illustration, courtesy Ford Europe Ltd)

Part III: ROAD TESTING

25. Never buy a car without first road-testing it, or getting someone who has a good feel for cars to road-test it Once you are on the road, you're looking for all the usual things such as a smooth clutch and gearchange, steering that

25

doesn't tug, brakes that don't pull, etc.

You should also check that acceleration is smooth, strong and stutter-free, and that the engine decelerates immediately upon release of the accelerator pedal. If it doesn't, this could be due to blocked fuel-return lines or to a sticky throttle cable or injection pump linkages.

Engine Surge

Some diesel cars occasionally suffer from a phenomenon known as "surging". With this condition the car momentarily accelerates independently of the accelerator pedal position. Watch out for this: it is commonly associated with air leaks into the fuel system. It can be put right, but it may take hours to find the point of leakage. If a car is good on all other counts, but surges, don't give up on it, just make sure that it is corrected before you buy.

Air ingress to the fuel system can also give rise to other symptoms, such as misfiring, lack of power, an inability to rev, or stalling and difficult re-starting.

One other point: on high-mileage diesel engines, surge may be due to excessive crankcase pressure causing engine oil to travel into the air intake and fuel the incoming air. Check for this as detailed under Crankcase Ventilation on *page 37.*

Over-cool Running

Keep your eye on the engine temperature gauge: if it never

gets to the half-way point, chances are the engine has been running too cool for a long time, and that's not good for its lubricating oil, fuel efficiency or for internal engine wear. Cool running may also be accompanied by deposits of white or beige sludge inside the oil filler cap, the crankcase breather hoses and on the dipstick. Avoid a car with these symptoms.

Overheating

Also watch the gauge for overheating, which can be even more damaging to the engine. Overheating is usually caused by ineffective cooling system operation, which may be the outcome of the presence of sludge and corrosion in the system, a dirty or obstructed radiator, air locks or insufficient coolant. Check the coolant level and look for the tell-tale signs of limescale and crystalline anti-freeze deposits around all coolant hose connections.

Other possible reasons for overheating are incorrect injection timing (which can be rectified cheaply), or malfunctioning injectors (considerably more expensive). If you have any reason to believe the engine may have run over-temperature for some time, avoid the car, because overheating may well have caused internal engine damage.

IMPORTANT NOTE: Severe damage will be caused if a diesel engine is run without coolant. Check all cooling and heater hoses because a sudden loss of coolant could be catastrophic!

Part IV: CHECK LIST

Take this list, or a copy of it, with you when you go to look at a vehicle, to make sure that you don't miss any of the major points mentioned in the chapter.

☐ 1. Insist on a thoroughly documented service history - unless the car is in the "cheapie" category, in which case you can't be too fussy!

☐ 2. Check the oil level, the colour of the engine oil, and for sludge in the oil and crankcase ventilation systems.

☐ 3. Look for oil in the air filter casing.

☐ 4. Check that the pre-heat warning light comes on, then goes out within 20 or so seconds.

☐ 5. Insist on starting the car from stone cold, and watch out for reluctance to start, rough idling and a dirty exhaust.

☐ 6. Listen for undue knocks or rattles from the engine, allowing for the characteristic "diesel knock" sound. A clattery cold engine should quieten considerably as it warms up.

☐ 7. Check exhaust colour on a hot engine. Rev the engine hard, watching for clouds of black, blue or grey/white smoke.

☐ 8. Check for engine oil leaks, particularly around the timing belt cover.

☐ 9. Make sure the car accelerates strongly and cleanly, and be wary of independent engine surge.

☐ 10. Where possible, have a cylinder compression check carried out on the engine.

CHAPTER 5 - DIESEL SERVICING

Everyone wants to own a car that starts first time, runs reliably and lasts longer than the average. Well, there's no magic about how to put your car into that category; it's all a question of thorough maintenance!

The diesel car owner is blessed in many ways in comparison with his petrol-car owning counterpart. His car offers a variety of major advantages which include greater fuel economy, greater durability and driving flexibility, the use of a safe, non-volatile fuel. And most importantly to some, routine engine maintenance that is well within reach of the majority of owners!

Due to the absence of an ignition system and carburettor, the diesel engine eliminates the need for fiddling with points, plugs, distributor cap, HT leads, coil connections and even the infamous carburettor mixture screw. Unlike petrol cars, diesels don't need regular 'tune-ups'.

On petrol engines the most persistent problems of non-starting, misfiring and stalling are nearly always due to the eagerness of high ignition voltages to escape inadequate insulation, especially in damp weather. The diesel engine suffers no such problems. However it's by no means immune from electrical problems associated with the charging and starting systems (starter motor circuit). These, incidentally, are identical to those of petrol cars, if a little more robust.

Despite the technical complexity of the diesel's fuel injection system, it requires comparatively little maintenance. Delicate engineering precision is confined to the internal workings of the injection pump and the injectors, which the DIYer must not - and will probably never need to - become involved with. Fortunately the injection pump - probably the single most expensive engine component - rarely gives any trouble, and usually outlives the engine itself.

The Only Disadvantage

The diesel's greatest weakness is a thirst for engine-oil changes: on many diesel cars these are required more frequently than on petrol engines, although many modern diesels can now make do with oil change intervals comparable to those of petrol engines. This really is the only maintenance disadvantage the diesel presents.

From the point of view of overall engine maintenance, there is nothing unusual about the air and oil filtration, drive belts or

valve tappets - nor any other aspect of the car, such as the transmission, suspension, brakes or chassis.

For these reasons, this chapter deals only with engine maintenance, and concentrates on aspects specific to the diesel, although you will find advice concerning topics - such as air filter renewal and valve clearance adjustment - that are common to both diesel and petrol engines.

How To Use This Chapter

The information in this chapter is of a generic nature, as this isn't a book about any one particular diesel car. The principle of everything we discuss here is pretty well the same from one diesel car to another, and by selecting the most popular underbonnet configurations and components as examples, we've covered the vast majority of eventualities. Using this in-depth diesel knowledge in conjunction with a car-specific manual (unavailable with engine information for the majority of diesels - we hasten to add!) and technical data, you will be able to guarantee a job well done.

Don't forget that we've included useful servicing data for the most popular diesel cars in *Chapter 8, Facts and Figures,* and make-by-make service schedules in the closing section of this book - the *Service History*. If you ensure that the Service Jobs discussed here are all carried out at the intervals specified by your car manufacturer, you can almost guarantee that your car will still be going strong when others have fallen by the wayside - particularly as it's a long-lived diesel!

For ultimate ease of use, each descriptive section is supported by illustrations - a photograph or a line drawing - usually on the same page, and with direct references in the text.

Special category qualifications **OPTIONAL** and **SPECIALIST SERVICE** are used next to some of the following sub-section headings. **SPECIALIST SERVICE** means we recommend you have this work carried out by a specialist. This is because some jobs, such as checking and adjusting the injection timing or setting the idle speed, can only be done with the right measuring equipment, which you may consider too expensive to be worth buying. Other jobs, such as injector overhaul and calibration, also demand the use of specialised equipment - and a certain amount of experience. Where we think you are better off having the work done for you, we say so!

INSIDE INFORMATION brings you all the useful hints and tips from the experts in the trade that help towards making the job easy - and right!

Each maintenance job is detailed within a section which relates to a particular mileage/time interval - e.g. Every 6,000 miles or 6 months. This does not mean that you must carry out that job on your car at this interval, rather, it is a typical interval for the job in question, and one which, if anything, errs on the side of caution. You should ideally use this servicing guide in conjunction with your vehicle manufacturer's service schedule.

making it easy! We think it's very important to keep things as straight-forward as possible, and where you see this heading you'll know there's an extra tip to help 'make it easy' for you!

The **OPTIONAL** reference means that you may wish to use your own discretion as to whether to carry out this particular job, as it isn't something that is normally specified on service schedules. Not all the operations discussed necessarily apply to any one car, so some can be skipped if your enthusiasm doesn't run that deep! A core few cannot be escaped though, and these are: oil and filter change, fuel filter water draining, fuel filter replacement and bleeding - the essential jobs which must be regularly carried out to prevent undue wear and tear of the engine, and the onset of unreliability.

For engine diagnostic advice, which includes glow plug, fuel system and cylinder compression testing, refer to *Chapter 6, Fault Finding,* later in this book.

The 'Catch-up' Service

When you first buy a used car, you never know for sure just how well it's been looked after.

So, if you want to catch-up on all the servicing that may have been neglected on your car, just work through the entire list of maintenance jobs discussed in this chapter, and your car will be bang up to date and serviced as well as you could

hope for. Do allow several days for all of this work, not least because it will almost certainly throw up a number of extra jobs - potential faults that have been lurking beneath the surface - all of which will need putting right before you can 'sign off' your car as being in tip-top condition.

The Service History

You may well want to keep a full record of all the work you have carried out, so, after servicing your car, you will be able to tick off the jobs you have completed. In this way you will build up a complete Service History of work carried out on your car, service by service.

Those people fortunate enough to own a new car, or one that has been well maintained from new will have the opportunity to keep a service record, or 'Service History' of their car, usually filled in by a main dealer. Until now, it hasn't been possible for the owner of an older car to keep a formal record of servicing, but now you can, using the complete tick list in *Appendix 3, Service History.* There's also the additional bonus that there is space for you to keep a record of all of those extra items that crop up from time to time.

New tyres; replacement exhaust; extra accessories; where can you show those on a regular service schedule? Now you can, so if your battery goes down only 11 months after buying it, you'll be able to look up where and when you bought it. All you'll have to do is remember to fill in your Service Schedule in the first place!

SAFETY FIRST!
*This is one more special heading you'll come across from time to time, and it is the most important of all! SAFETY FIRST! information must always be read with care and always taken seriously. In addition, please read the whole of **Chapter 1, Safety First!** before carrying out any work on your car. There are many hazards associated with working on a car but all of them can be avoided by adhering strictly to the safety rules. Don't skimp on safety!*

Every 500 Miles, Weekly, or Before Long Journeys, Whichever Comes First

Engine Bay

☐ **Job 1. Engine oil level.**

Although some engines barely need their sump oil topping-up between major services, even healthy ones sometimes have an unusual appetite for it, while worn ones will certainly burn it. New or old can develop an oil leak. An engine low on oil runs the risk of internal damage, over-heating and eventual seizure - all of them ruining the engine.

Before you check the oil level, the engine should be switched off and left standing for a while to ensure that all oil has returned to the sump - obviously first thing in the morning after garaging overnight is the ideal time.

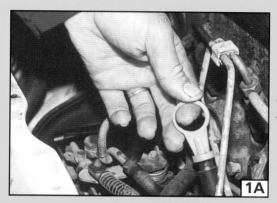

1A. Obviously, too, the car should be on level ground when you check the dipstick. Your handbook should identify the stick's location, but if not, look for a 'ring pull' disappearing into a tube or hole on the side of the engine.

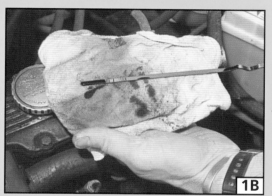

1B. Withdraw the stick and wipe it dry on lint-free cloth. There will usually be MAXIMUM and MINIMUM marks on the stick and some handbooks will quote a specific quantity of oil for the difference between the two (often one litre).

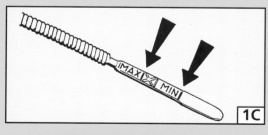

1C. Reinsert the dipstick then withdraw it again - the oil level will be seen on the dipstick and should be on or pretty close to the MAXIMUM (or MAX) mark, not close to the MINIMUM (or MIN) mark. The lower mark is, of course, the danger level. But it could also be unwise to exceed the maximum mark - a shade over won't matter, but substantial over-filling can also lead to over-heating and other problems. (Illustration, courtesy Ford Europe Ltd)

INSIDE INFORMATION: If your car's dipstick has to snake along a curved tube, and especially if the oil is fresh and clear, you might have difficulty in seeing the oil level. Dip, wipe and re-dip several times, turning the stick so that it goes into the tube from different angles, if possible - sometimes oil is wiped on or off the dipstick as it is withdrawn and replaced, but only on one side. If it's the side with the markings on, put the stick in the other way round.

1D. If you do need to top up, do so a little at a time, through the oil filler normally found on the valve cover at the top of the engine. Allow time for the new oil to reach the sump before you re-check the level, not forgetting to start with a clean dipstick again.

1E. Most oil filler caps unscrew 1/4 turn and lift off.

1F. A few screw all the way in and out. Don't force them down too tight when refitting.

☐ **Job 2. Coolant level.**

SAFETY FIRST!
i) The coolant level should be checked WHEN THE SYSTEM IS COLD. If you remove the pressure cap when the engine is hot, the release of pressure can cause the water in the cooling system to boil and spurt several feet in the air, with the risk of severe scalding.
ii) Take precautions to prevent anti-freeze coming in contact with the skin or eyes. If this should happen, rinse immediately with plenty of water.

Again, this is a check that should be made first thing in the morning, before the engine has been run. There are two reasons for this: (i) Hot coolant expands, so you'll only get a true level reading when it's cold; (ii) Hot coolant (like a boiling kettle) can be extremely dangerous, and removal of the filler cap can release a scalding blast of steam and liquid. See *SAFETY FIRST!* above, and *Chapter 1*.

2A. NEVER ATTEMPT TO REMOVE THE FILLER CAP WHEN THE ENGINE IS HOT. If, in an emergency, the cap needs to be removed before the engine has completely cooled, wrap a rag around both the cap and your hands, and open the cap in two stages, the first quarter-turn to release any remaining internal pressure. Generally speaking, an engine is at its most efficient when running at a temperature close to that of boiling water - and generally speaking, it is water, with an anti-freeze content, that is used for the coolant. And water under pressure can reach a higher temperature than normal without boiling, thus avoiding all the attendant dangers of its dissipating away in steam, or even exploding the system!

2B. The integral valve mechanism of a pressure cap will allow pressure within the system to build to a pre-determined level, say, 7 or 13 psi. At whichever point, a valve in the cap is forced open, allowing further pressure to vent to atmosphere (older systems) or (modern systems, as shown here) letting the expanding coolant out of the main radiator and into the expansion tank. As the engine cools, so does the coolant, which then begins to contract, causing a vacuum to start building within the system. This sucks open another valve in the cap, the suction then drawing (respectively) either air back in, or, with modern systems, the previously expelled coolant back from the expansion tank.

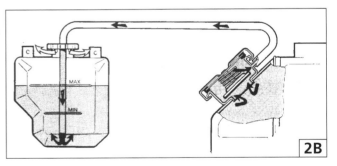

2C. When cold, the coolant level in an old system should be just below the radiator filler neck...

2D. ...or up to the level mark scribed on the expansion tank of a modern system. Where necessary, top up with clean water mixed with the percentage of anti-freeze (e.g., 25%, 33% or 50%) as recommended in your handbook.

Job 3. Battery electrolyte.

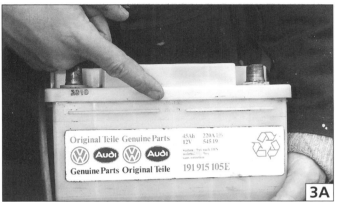

Where the battery is provided with screw caps to the individual cells, or obviously removable strips which plug into or over a number of cells at a time, it is obviously intended that its electrolyte content should be topped up as and when required. But note that even a so-called 'maintenance-free' battery may have flush-fitting strips over its cells which can be prised up for the addition of electrolyte, perhaps prolonging its life beyond general expectation! (Not all do, so don't force it!)

3A. If the battery case is translucent, look for a level mark scribed on its side, otherwise a general recommendation is that the electrolyte level should be just above the tops of the plates which you can see with the cell caps or strips removed.

3B. Some modern batteries have a 'magic eye' which changes colour to indicate the battery's state of charge, or to warn that it is low on fluid.

INSIDE INFORMATION: Note that here is an instance where it is preferable that the battery be warm, as after a run, before checking the level, since the electrolyte expands with heat. If it were topped up while cold there is a danger that later the fluid would overflow, leading to corrosion of the terminals.

If checking now reveals corroded terminals (typically, a white, powdery growth) refer to *Job 21. Battery terminals* in the *12,000-mile/12-month Service* section.

3C. Top up only with distilled (de-ionised) water, never ordinary tap water, which may contain impurities which would damage the plates and shorten the battery's life. Mop up any accidental spillage immediately, and make sure the entire battery exterior is clean and dry. You can buy de-ionised water from the auto accessory shop, either in handy top-up bottles or in bulk containers - note that it is highly recommended, too, for steam irons!

3D. Here's how to check the strength, or specific gravity, of the battery electrolyte. You place the end of a hydrometer into the battery electrolyte, squeeze and release the rubber bulb so that a little of the electrolyte is drawn up into the transparent tube and the float or floats

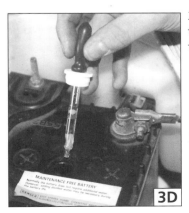

inside the tube (small coloured beads are sometimes used) give the specific gravity. Check each cell and if one (or more) is significantly lower than the others, with battery topped up, the battery is probably on its way out.

3E. Check the tightness of the battery clamp. A loose, rattling battery will have a shorter life than one that is held down securely.

3,000 Miles - or Every Three Months, Whichever Comes First

The Engine Bay

☐ **Job 4. Generator drive belt.**

SAFETY FIRST!
Always keep your fingers away from the blades of a thermo-electric cooling system fan unless the battery has been disconnected - it could surprise you by coming on suddenly when the engine is running, and there are some which are not ignition circuit controlled and which will certainly surprise you by coming on (perhaps soon after parking up) when the engine has been switched off!

Still often referred to as the 'fan belt' because on older cars it drove the mechanical cooling fan (usually attached to the water pump) as well as the generator (a dynamo, or later, an alternator) this drive belt continues to be a vital component. On a modern car, while the fan is usually thermostatically controlled and driven by an electric motor, the belt probably still drives the water pump, as well as the alternator. If it slips, or breaks, the car will eventually grind to a halt, having run out of electricity, or overheated, or both.

4A. Even though the diesel engine doesn't require electricity to operate an ignition system, it does require a continual voltage to the fuel pump shut-off solenoid, which otherwise closes under spring-pressure, shutting off fuel supply to the pump. Also, modern 'high-tech' diesels, such as the Volkswagen-Audi TDIs, now have an electronically controlled fuel injection system.

4B. In order to work efficiently, as well as being in good condition the belt must be reasonably tight - but not over-tight - around the drive pulleys. Look on the inside of the belt for a shiny surface, cracks or oil; look at the edges for fraying. Replace if there are any problems.

4B. Checking belt condition can be awkward, both from the point of view of restricted access and the difficulty of seeing enough of it in its fitted state: you really need to be able to rotate the engine (ignition off) using a spanner on the alternator or crankshaft pulley nut, bending and twisting each newly exposed length of belt to show up cracks or other damage.

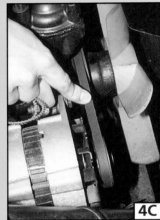

4C. For the belt tension check, a typical recommendation is that there should be approximately 10 mm (1/2 in) deflection, (i.e. 5mm in each direction) but check with your car's handbook. Do not overdo the belt tension, for this puts undue strain on the generator and water pump bearings. Test with firm thumb pressure midway on the belt's longest run between pulleys.

4D. If adjustment is required, the generator pivot bolt(s) and slotted adjustment strap bolts, on engine and generator, need to be slackened. The generator is then pivoted away from the engine until the belt is sufficiently taut, and the bolts then re-tightened.

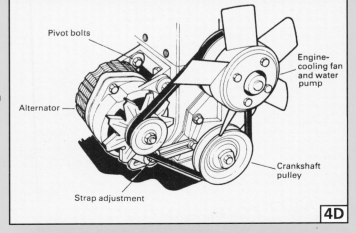

Pivot bolts

Engine-cooling fan and water pump

Alternator

Crankshaft pulley

Strap adjustment

3,000 MILE SERVICE

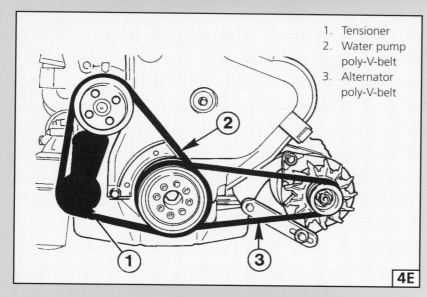

1. Tensioner
2. Water pump poly-V-belt
3. Alternator poly-V-belt

4E. There will often be small variations on the theme. For instance, the Sierra turbodiesel has a larger water pump than standard, with its own drive belt (2) and idler pulley (1). The principles of adjusting the belt (3) by slackening the adjuster bolts, are exactly the same. (Illustration, courtesy Ford Europe Ltd)

4E

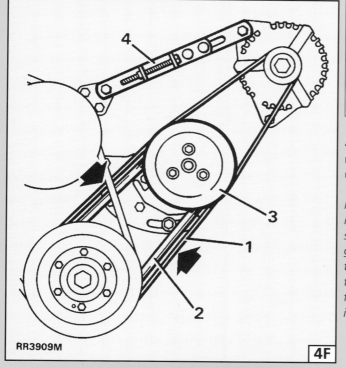

4F

RR3909M

making it easy! If necessary, a suitable piece of wood can used, carefully, as a lever between the engine and the generator drive end bracket to maintain the belt tension while the bolts are tightened. But don't **over** tighten the belt - it will damage the alternator - and take care not to crack the alternator casing.

4F. Very occasionally, you may come across a simple threaded rod and bracket tension adjustment (4F.4). Simply tighten or undo the rod, as required.

INSIDE INFORMATION: In view of this drive belt's vital importance, you should always carry a spare and the right sized spanners to slacken the adjustment bolts: if, even when the generator is pushed tight to the engine, the new belt is difficult to get over the pulleys, follow the old bicycle chain and sprocket trick - push as much of the belt as possible onto the edge of the final pulley, then rotate this pulley to wind the rest of the belt into place.

Around the Car

☐ **Job 5. Pipes and hoses.**

5. Carry out a visual check on all flexible and rigid pipes and hoses in and around the engine bay for leaks.

Under the Car

☐ **Job 6. Inspect for leaks.**

While under the car, look out for fluid leaks, such as hydraulic fluid spotted on a tyre wall, or oil dripping from beneath the engine or transmission. It is better to spot such leaks early, before danger threatens or major expense is required. If brake fluid leaks are found, don't use the car until the problem is resolved!

5

6000-mile/6-month Service

Carry out the 3000-mile/3-month service operations, plus the following:

The Engine Bay

FACT FILE: FUEL FILTER

Water can be very damaging to the internal components of the injection pump and injectors, as if it is left in contact with the finely machined surfaces of these components it causes corrosion, subsequent wear, and loss of the exacting operating tolerances required for accurate fuel metering and high-pressure injection. Unfortunately, water is ever-present in diesel fuel, and especially so in low-quality fuel bought in some countries. Even if the fuel quality is high, water enters it due to natural condensation occurring in filling station storage tanks, and the vehicle's own fuel tank. The fuller the car's tank is for most of the time, the less likelihood of the diesel fuel becoming contaminated with water.

1. A water trap or sedimenter is included in-line with the fuel supply to the injection pump, usually as an integral part of the fuel filter. It is in the form of a bowl below the filter element, fitted with a drain orifice and screw-plug at its lowest point. Not only does it collect water (which is heavier than diesel) but also the larger pieces of contaminant which may find their way into the system - such as rust from the fuel tank. Water draining is usually specified at every service, and it should be done at least every 6,000 miles or 6 months.

2. If your car is fitted with a water level sensor in the filter though, wired to a tell-tale lamp at the facia panel, you can afford to leave water-draining for longer. Cars with a Lucas filter but no such sensor, can have one retrospectively fitted.

There are three main designs of diesel fuel filter. Pic 3. shows a removable canister (A) with a disposable inner element (B) and sealing ring (C); Pic 4. has a disposable filter body (A) sandwiched between water bowl (B) and filter head (A). Pic 5. is a fully disposable filter cartridge (A), being attached to the pump and oil cooler connections on this Ford Unit.

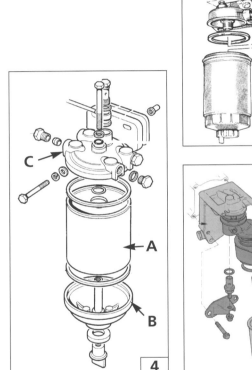

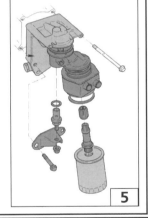

☐ Job 7. Draining water from fuel filter

SAFETY FIRST!
Whenever you are dealing with diesel fuel, it's essential to protect your hands by wearing plastic gloves.

7A. Draining is achieved simply by undoing the drain screw by several turns (it may be necessary to remove it completely on some makes of filter) and refitting/tightening it when fuel emerges from the filter free of water. Water can be seen as large, colourless bubbles in the fuel. If you're not sure of what you're looking at, just evacuate about 100 cc, or a cupful of fluid and you can be sure you've rid yourself of any trapped water.

7B

making it easy! With some filter types you may find that fluid does not emerge when the drain tap is opened. This is due to the negative-pressure operation of the fuel supply system (in other words, the fuel/water is retained in the filter by suction). If this is the case, you can break this vacuum by opening the vent screw at the filter head, or, if no such screw is present, by slackening the fuel outlet union at the head (the union of the fuel hose which connects with the injection pump). After opening the system in this way, it may be necessary to bleed the fuel system of air if you encounter engine-starting difficulties. System bleeding is detailed in *Job 20,* under the *12,000-mile/12-month* section.

7B. Because diesel fuel is damaging to rubber components such as cooling system hoses and drive belts, it may be helpful to push a length of plastic piping onto the drain tap (nearly all have a stub for this purpose) to allow direction of the fluid into a suitable container. Some diesel cars are equipped as standard with a drain hose.

7C

7C. Peugeot-Citroen cars powered by the 1.7/1.9-litre XUD engine and produced from 1992 are equipped with a unique type of fuel filter which has its drain stub and tap pointing forwards or upwards (arrowed) from the base of the filter assembly. The filter itself is usually mounted on a casting which protrudes from the front of the cylinder head.

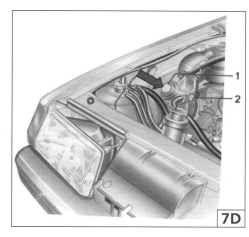

7D

7E

7D. This Citroen BX filter, however, is on the inner wing (arrowed). Drain by loosening the drain plug (2) then pumping the plunger (1) until no more water comes out.

7E. The Purflux filter fitted to some models is drained in the same way. (Illlustrations, courtesy Peugeot)

8

☐ Job 8. Accelerator controls.

8. Lubricate the accelerator control linkage at the injection pump, and (where applicable) the throttle pedal pivot, in the recesses of the footwell. Use spray-on lubricant or white silicone grease in the footwell, so as not to spoil your shoes with dripping oil - it stains leather!

☐ Job 9. Coolant system check.

9A. Check the cooling system for leaks and all hoses for condition and tightness. Look at the ends of hoses for leaks - check clamps for tightness and for cutting through the hoses. Pinch the hoses to ensure that they are not starting to crack and deteriorate. If you don't want a hose to burst and let you down in the worst possible place, change any hose that seems at all suspect.

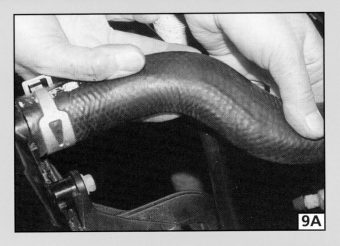

9B. Rather then employ too much force, particularly on a radiator, where there's a high risk of fracturing the hose stub, simply slide the clip out of the way and use a strong, sharp knife to carefully slit the hose until you can open it up and peel it off the stub.

| *making it easy!* | A spot of water on the rubber, preferably with an added drop of washing-up liquid, eases the cutting. |

Thoroughly clean the stubs, carefully using a file and emery cloth to remove the lumpy corrosion often found on elderly alloy cooling system components.

Position new clips (preferably of the flat, worm-drive type) on the new hose, ensuring their tightening screws are best placed for easy screwdriver access when the hose is fitted. A smear of washing-up liquid will help the hose slide fully home on the stubs. Tighten the clips firmly, but don't bury them in the hose.

9C. It is also worth checking to ensure that an accumulation of autumnal leaves and dead bugs is not restricting the airflow through the radiator.

☐ Job 10. Coolant check.

SAFETY FIRST!
i) Only remove the pressure cap when the engine is cold.
ii) Take precautions to prevent anti-freeze coming into contact with the skin or eyes. If this should happen, immediately rinse with copious amounts of water - seek medical advice if necessary.
iii) Never work on or touch the cooling system while the engine is running.

10. Use an anti-freeze hydrometer (available quite cheaply from accessory shops) to check the specific gravity (anti-freeze content) of the coolant. The tester will probably show a reading using coloured balls. If the concentration is below the recommended amount, top up the system with anti-freeze until the correct specific gravity is obtained. Of course the engine will have to be run for the newly introduced anti-freeze to mix thoroughly otherwise a false reading will be obtained. If you have any doubt over the period that the old mix has been in the car, drain and refill it with fresh.

INSIDE INFORMATION: Some owners think that there is little to be gained by using anti-freeze all the year round, particularly in those parts of the world where frost is not a problem. Wrong! Anti-freeze even in the minimum concentration of 25% not only gives protection against around -13 degrees Centigrade (9 degrees Fahrenheit) or frost, it also contains corrosion inhibitors to stop the radiator from clogging and so helps to keep the car running cooler.

☐ Job 11. Check water pump.

SAFETY FIRST!
Ensure the engine is turned off and the battery is disconnected before examining the water pump..

11A. Check the water pump for leaks - the first sign of mechanical failure - by looking for water leaks or stains around the spindle. Many pumps are so difficult to get at, particularly as some are driven by the camshaft drive belt, that it will be almost impossible to do this visually, so the best you can do is to feel around the pump and check with your hands for water or try rocking the cooling fan, if it is of the mechanical type.

SPECIALIST SERVICE. Any problems should be dealt with by your specialist if much has to be dismantled to get access to the pump.

☐ Job 12. Check power steering fluid.

If your car has power assisted steering, it will almost certainly have an oil reservoir.

12A. Unscrewing the lid (first clean away any surface dirt) should bring with it an integral dipstick, marked with maximum and minimum levels. Sometimes the reservoir may be translucent, with the levels marked on its exterior.

12B. If not, you'll have to rely on the integral dipstick markings.

Note that these level marks may be duplicated for 'Hot' or 'Cold' conditions: the former might apply if, for instance, the car has just been driven for 20 minutes or so at around 50mph, the latter if the engine has not been run for about five hours. Beware of burns if you are making a 'Hot' check. Top-up as necessary, using the oil specified in your handbook - generally automatic transmission fluid (ATF).

☐ Job 13. Check clutch adjustment.

CABLE CLUTCH MODELS ONLY
This is a job that does not apply to all diesel cars, since manual clutch adjustment is becoming something of a rarity. Many modern cars have a built-in automatic adjustment on their clutch cable linkage, while hydraulic clutch release mechanisms are largely self-adjusting.

13A. Clutch adjustment is the elimination of unwanted free play in the clutch linkage: wear and tear here can reach the point where full clutch pedal movement barely releases the clutch - a fact underlined by difficult and noisy gearchanges. Greasing the cable and mechanism can make the clutch smoother to use.

Note, however, that generally a specified clutch adjustment would maintain measurable free play. Without free play wear of the clutch release bearing is exacerbated.

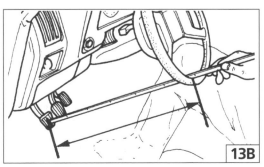

13B. Sometimes an adjustment specified in the car handbook (perhaps that the clutch and brake pedals should be level, or that the clutch pedal should be a certain distance from the floor or steering wheel) ensures that the bearing is in constant contact with the clutch. Obviously, the handbook for your car will tell you whether your clutch is adjustable, and if so, what the adjustment should be. (Illustration, courtesy Vauxhalll Motors)

13C. Typically, on an older, rear-wheel-drive car there is an adjuster where the linkage from the pedal meets the clutch release lever (or 'fork') at the gearbox clutch housing (or bellhousing). There is a threaded section with a nut and lock-nut. The adjustment specification is typically that, with the release lever (return spring temporarily removed) pulled away from the nuts until the release bearing is felt to contact the clutch, there is a gap of, say, an inch between the lever and nuts: if not, the nuts should be 'unlocked' and screwed along to obtain the correct gap - the specified free play.

13D. Sometimes a similar linkage and adjuster can be found on the transmission housing under the bonnet of early front-wheel-drive cars.

13E. Sometimes, instead, some sort of adjuster can be found where the clutch release cable passes through the engine compartment bulkhead. Here, rather than a threaded adjuster, there could simply be a C-clip in a choice of grooves - typically it might be specified that there should be five or six grooves between the clip and its abutment point when the outer cable is pulled away from the bulkhead.

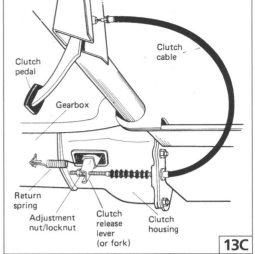

13C

HYDRAULICALLY OPERATED CLUTCHES ONLY
There were some older hydraulic-clutch cars that also had a threaded adjustment on the pushrod between the clutch slave cylinder and the clutch release lever. Generally, though, no adjustment is fitted (or required) on a hydraulic set-up.

Engine Bay/Under the Car

☐ **Job 14. Change engine oil.**

An engine oil change is typically recommended every 5,000 to 6,000 miles on diesels, and even as infrequently as 10,000 miles on some modern ones. Only a handful of years ago it was common to see 3,000-mile oil-change recommendations for diesels.

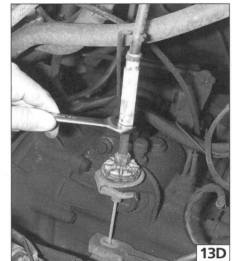

13D

One reason diesel engines tend to need more frequent oil changes than petrol engines is that the oil takes a greater molecular pounding, thanks to the enormous pressures of diesel combustion, and so doesn't maintain its ideal viscosity for as long. Another reason is that diesel fuel by-passing the piston rings and ending up in the sump
(just as petrol does) dilutes the oil but does not evaporate out because it isn't volatile. Oil dilution, of course, causes decreased protection.

Yet another reason for the frequency of oil-change intervals is that the by-products of diesel combustion are more corrosive than those of petrol, and more carbon is produced. If you add a turbocharger to the equation, local oil temperatures increase significantly, and this can lead to chemical changes in the lubricant and the deposition of harmful lacquers within the engine.

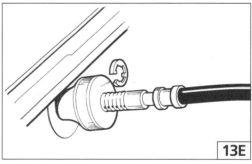

13E

It makes excellent sense to err on the side of caution and treat the recommended change interval as an absolute maximum, although users of synthetic oil can afford to stretch intervals slightly.

INSIDE INFORMATION: We would advise owners of turbocharged diesels to use only synthetic engine oil, as this is more resistant to the high temperatures reached inside a turbocharger, and it doesn't form lacquer deposits on hot bearing surfaces.

Probably the one service operation on which most DIY motorists will 'cut their teeth' is the engine oil change. No matter how technically complex the engine, an oil change still remains basically a simple operation.

Usually at the same time the engine oil filter will also be renewed, although a few service schedules stipulate that the filter needs to be changed only at every other oil change.

SAFETY FIRST!

All used engine oil is carcinogenic (cancer inducing) but used oil from a diesel engine is even more so; this is why it is essential to work cleanly, to protect hands and wrists with barrier cream and polythene protective gloves, and to regularly wash clothing which has been soiled with old oil. Refer to Chapter 1, Safety First!, for more information on used diesel engine oil and page 7 for details of your nearest oil disposal point.

Oil drain plugs are often overtightened. i) Take care that the spanner does not slip causing injury to hand or head. (Use a socket or ring spanner - never an open-ended spanner - with as little offset as possible, so that the spanner is near to the line of the bolt.) ii) Ensure that your spanner is positioned so that you pull downwards, if at all possible. Take great care that the effort needed to undo the drain plug does not cause the vehicle to fall on you or to slide off ramps - remember those wheel chocks. iii) Refer to the information in Chapter 1, Safety First!

14A. Prepare for the job beforehand, making sure you have both the oil (of correct grade and quantity) and - if you're going to change it - the filter. You'll also need a plentiful supply of clean, lint-free rag.

Make sure you have a suitable receptacle into which the oil can be drained: a washing-up bowl would do, but it has to be usable without having to raise the car. In other words, the bowl needs to be shallow enough to fit comfortably underneath the engine.

14B. Check whether the sump plug has a hexagon head which can be undone with a conventional ring spanner, or whether (perhaps recessed) it requires a special drain plug key. Do not attempt to make do or you risk rounding the drain plug, making it extremely difficult to grip with any tool.

14C. If you are going to have to raise the car to obtain sufficient working room, it will (as stressed earlier) need to be supported securely on ramps or axle stands: make sure you have these to hand.

making it easy!

i) When all preparations have been made, note that the best time to actually drain the oil is after a short drive, when the oil will be warm and will flow more easily. ii) Before undoing the drain plug, take off the oil filler cap: this relieves any partial vacuum in the system - the faster the oil can drain from the sump the more debris it will drag out with it. iii) If the drain plug is fitted into the side of the sump, bear in mind that the initial spurt of draining oil may carry some distance sideways. Position the drain tray (or bowl) accordingly, being ready to move it inwards as the spurt wanes and the oil falls vertically.

14D. Undo the last few turns of the drain plug by hand, giving yourself room to withdraw hand and plug quickly before the oil runs under your sleeve and all down your arm! Incidentally, check, by feeling the sump, that your short drive hasn't made the oil too hot to handle. Don't take any risks; let it cool a bit.

14E. If the drain plug falls into the bowl (as also often happens) don't worry, you can retrieve it later. When you do, clean it up thoroughly and (if it's not a taper-seat type) renew its sealing washer.

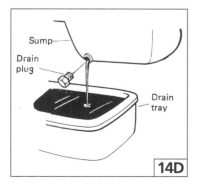

Sump

Drain plug

Drain tray

When the sump has dripped its last drop, you can clean up the sealing face of the sump drain plug and refit the cleaned-up plug - it should be done up tight, but not 'murdered'!

(If you are going to change the oil filter at the same time as renewing the oil, now refer to *Job 15* before continuing).

14F. Back up top, pour in the correct grade and quantity of fresh oil: do so slowly, as some engines have a nasty habit of spewing back, leaving oil running all over your clean engine, and reaching parts that you, with your piece of rag, cannot!

14G. Having checked that the oil level corresponds with the correct mark on the dipstick, start the engine and run it for a few minutes while you check around (basically at the filter unit and sump plug) for leaks. Satisfied that all is well, switch off and allow time for the oil to settle before re-checking the level, topping-up if necessary.

14F

Pour the old oil into an empty can, ready for eventual carriage to your nearest disposal point (probably at your local tip). Clean out the drain tray, clean up and put away your tools, dispose of dirty rags, and the empty filter box, etc.

SAFETY FIRST!
DON'T pour the old oil down the drain - it's both illegal and irresponsible. Your local council waste disposal site (see page 7) will have special facilities for disposing of it safely. Moreover, don't mix anything else with it, as this will prevent it from being recycled.

14G

☐ Job 15. Renew oil filter.

While the oil is draining, you can get on with tackling the oil filter. Fitted on the side of the engine somewhere, sometimes low down, sometimes higher up (often pretty inaccessible!) on an old car it might be of the bowl-and-element type and on newer cars a one-piece canister unit.

Before tackling the filter, however, check first that your drain tray extends far enough to catch any oil spillage as it is loosened, and if it doesn't, place another suitable, small receptacle beneath.

On the face of it, the throw-away one-piece canister should be the most straightforward to deal with. But often the bowl-and-element type, secured to the engine with a long through-bolt with conventional hexagon head, is the easiest to remove.

15A

15A. With the latter, having unscrewed the through-bolt, you then lift the bowl away, preferably bringing with the old one the receptacle beneath to contain any dripping oil: the bolt should remain captive within the bowl. Upending the bowl should tip out the old cartridge filter element. Watch out for a metal base plate and spring beneath the element, restraining them should they try to slide down the bolt, although they should, in fact, also be captive. Discard the old filter element and rinse out the bowl with paraffin, using a brush, if necessary, then dry the bowl inside and out with clean, lint-free cloth.

15B. Inside the box containing your new filter element should also be a narrow sealing ring. You'll probably find the old one still stuck in the groove in the filter head on the engine (the part to which the bowl was bolted) - it has to be carefully prised out with the point of a dart or a slim screwdriver blade.

Once it's out, (and it's usually pretty inaccessible, or at least difficult to see!) make sure the groove and the face of the filter head are thoroughly clean and dry.

Now, bearing in mind how well stuck in was the old seal, you'll be surprised how difficult it can be to get the new seal to stay in place while you offer up the bowl and new filter, working against spring pressure to ensure the narrow rim of the bowl is seated evenly on the seal while you screw in and partially tighten the through-bolt.

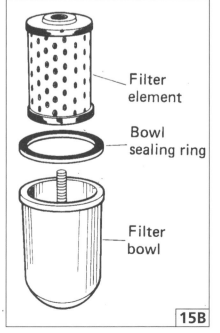

Filter element

Bowl sealing ring

Filter bowl

15B

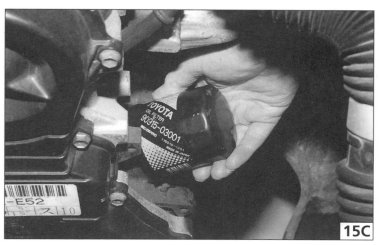

INSIDE INFORMATION: Smearing the seal with grease before positioning it will help to retain it, but you must make sure now that it hasn't 'escaped' anywhere before you fully tighten up the bowl - tight, but not 'murdered'!

15C. The modern screw-on, screw-off, throw-away canister oil filter should be a much easier proposition. And it is - provided you can unscrew it! Although when fitted it is (or should be) only screwed on hand-tight, it is amazing how tightly it sets in service. The problem is compounded when, as is so often the case, the filter is inaccessible.

If access is that bad, but you can get your hands to it, one trick that often works is to wipe clean the outside of the filter wrap abrasive paper around it, grip it tightly and unscrew.

15D. Hopefully, though, there will be enough room to attack it with one of the various DIY filter straps, or wrenches that are on the market.

INSIDE INFORMATION: If all else fails, drive a screwdriver right through the filter and twist it loose, using the screwdriver as a lever/handle.

15E. Once off, it's simply a case of cleaning up its mounting point, moistening the captive seal of the new canister with clean oil, then screwing on the new filter - hand tight only!

12,000-mile/12-month Service

Carry out the 6000-mile/6-month service operations, plus the following:

The Engine Bay

☐ Job 16. Glow plug maintenance.

Rarely will you see any mention of the glow plugs or pre-heating system made on service schedules. This is because manufacturers regard the cold-starting system as one that either works or doesn't... and if it doesn't, you can probably set things right by renewing all the glow plugs.

However, as a keen DIYer you'll no doubt want to keep an eye on the condition of the plugs, which can be a handy indicator of injector problems. Our recommendation is that you remove all glow plugs at each major service (typically the annual service), wipe the soot from them, check there's no erosion, and refit them. It's best to catch worn-out plugs before you start experiencing starting difficulties.

GLOW PLUG REMOVAL

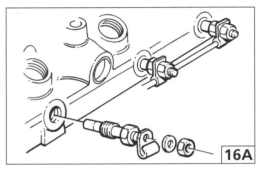

16A. Before removing a glow plug from the cylinder head, disconnect the battery earth lead. It is important that you first refer to *FACTFILE: COMPUTER PROTECTION,* under *Job 21.* Battery Terminals, later, to avoid damaging electronic components. Next, disconnect the wire or connecting strap from the plug and unscrew it from the cylinder head by just a couple of turns using a ring spanner or socket. (Illustration, courtesy Ford Europe Ltd)

16B. Clean away dirt from around the plug so that none finds its way into the engine once the plug is removed, then fully unscrew the plug (and its sealing washer, if fitted). It's a good idea to blank off the plug hole with a piece of cloth to prevent dirt from entering.

REFITTING THE GLOW PLUG

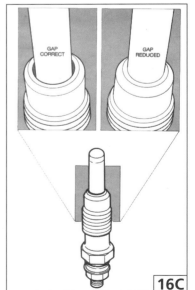

16C. Where a plug sealing washer is fitted this should be renewed before refitting the plug. If a plug tightening torque is specified in your model-specific manual or handbook, it is important to observe it, as overtightening can cause damage. Admittedly, it isn't very easy gauging a tightening torque when you're using a spanner; the important thing is not to over-tighten, as this can damage the plugs. This is because an annular gap between the screw body and the element sheath will close up if overtightened, so altering the plug's heating characteristics, causing it to draw excessive current and burn out all the sooner. Make sure that the electrical connection to each plug is clean and corrosion-free, then refit the supply wire or strap before reconnecting the battery. (Illustration, courtesy Hella)

☐ Job 17. Check valve clearances.

Many modern diesel engines have hydraulic tappets whereby the valve operating mechanism is controlled by oil pressure so that it can automatically compensate for any play developing between the working parts (whether through mechanical wear or heat expansion). On older engines there is built-in play and a means of adjusting it as required.

The valve clearance is a pre-determined gap that is essential to the health and performance

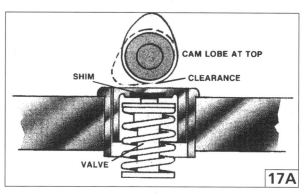

17A

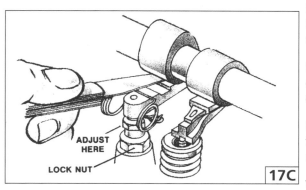

17B

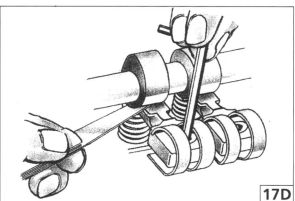

17C

17D

of the engine. It is a gap maintained between the valve stem (when the valve is closed) and the means employed to open the valve in order to let the fuel mixture in (inlet valve) or to let out the waste gas (exhaust valve).

In order for the combustion and the power stroke of the piston to be as effective as possible, it is essential that during these phases the valves are firmly closed.

If there were firm contact at all times between the valve stem and rocker arm or camshaft lobe there is a danger that valves could be held partially open when they are supposed to be shut. A gap is essential to allow for expansion of components as the engine heats up.

But at the same time there must not be too large a gap, or the valve will not fully open. This is why a critical gap - the valve clearance - must be maintained.

Make sure, when you look up the gap specification in your car data, that you note whether the setting is measured with the engine cold (e.g., after standing overnight) or hot (i.e., at normal running temperature).

17A. On many overhead camshaft (ohc) engines (the majority of car diesel engines) where the camshaft is mounted in the cylinder head directly above the valves, valve operation is directly via the camshaft lobes, and the gap is controlled (typically) by specifically sized shims in buckets over the valve stems. Here, gap adjustment, when required, is achieved by adding or subtracting various thicknesses of shim.

SPECIALIST SERVICE: Often this adjustment demands a degree of expertise and use of expensive special equipment that rather puts the operation beyond DIY capability.

Fortunately, the shim-type arrangement has fewer working parts, and thus less scope for wear and tear, the need for adjustment rarely arises until comparatively high mileages - usually well above sixty thousand miles.

Where the engine is an overhead valve (ohv) or pushrod unit, the camshaft is mounted lower down in the engine rather than in the cylinder head itself, and operates the valves via a rod and rocker arm mechanism. (Illustration, courtesy Gunsons)

17B. The rocker arm (the bit that actually bears against the valve stem) incorporates a screw adjustment at the point of contact with the pushrod, which makes gap-setting fairly straightforward. There are as many rocker arms (either sharing a common rocker shaft or set on individual pedestals) as there are valves. (Illustration, courtesy Gunsons)

17C. There are some ohc engines in which the camshaft lies alongside the valves in the cylinder head, with the shaft also employing rocker arms between itself and the valves.

Here the valve gap adjustment is almost as straightforward as that for the ohv pushrod units, except that often access is rather restricted. Sometimes, too, the gap may be measured between the cam lobe and the rocker, rather than between the valve and the rocker, while occasionally measurements for both are given. Since the mechanism is still relatively compact, the need for adjustment again only occurs after a high mileage.

Often the specified gap is the same for both inlet and exhaust valves, typically 0.012 in. (0.305 mm), which helps simplify matters. But sometimes there are two different measurements, say, 0.008 in. (0.22 mm) for the inlets and 0.023 in. (0.59 mm) for the exhausts - so here you need to know your exhausts from your inlets - and your manufacturer's recommendations!

17D. Some VW and Audi engines require the use of a 7mm Allen key for adjusting the gaps.

17E. Obviously, the identification of valve positions is best done with the valves exposed and, if indeed, your engine's valve adjustments need to be at least checked, if not (by dint of their apparent noisiness) actually requiring adjustment, then to gain access to them the valve cover must be removed. Be prepared to renew the valve cover gasket shown here.

Since a valve clearance must only be checked or adjusted when the valve is closed, you need to rotate the engine to ensure that this is so.

making it easy! 17F. This is easier to do if the glow plugs or injectors are first removed to relieve cylinder compression. In fact, given the diesel's compression ratio, it is virtually impossible to rotate the engine without first removing them. You can turn the engine either by jacking up one driving wheel and turning the wheel with fourth gear selected, or by using a spanner on the crankshaft pulley nut or on the generator pulley nut - you may have to carefully lean on the drive belt to prevent it slipping.

INSIDE INFORMATION: Note that while most engines turn clockwise in their normal direction of rotation, some - notably a few Japanese units - run anti-clockwise, so you'll need to check this beforehand.

As we said, checking or adjustment demands that the valve in question be fully closed. There are various ways of ensuring this, some involving useful 'short cuts' evolved by mechanics over the years. These can save you time, or you may have to go the long way round.

Of course, if it's an ohc engine - as most car diesels are - you can see the camshaft - and you'll be able to see that where the peak of a camshaft lobe points directly away from the valve or rocker, the valve is fully closed.

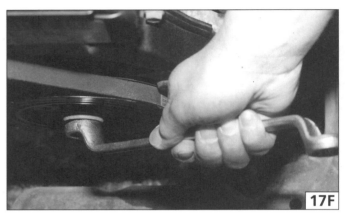

On the 'pushrod' ohv engine you can't see the camshaft lobes but you can see the rockers and valves, and it is these that must be observed to determine whether valves are open or closed. Usually it is easiest, while rotating the engine, to watch a rocker arm bearing down on the valve and pushing it fully open: if you turn too far and the rocker starts to come up again, simply turn back and then forth until you've got it right.

making it easy! The commonest 'short cut' which can be applied to many four-cylinder engines with a (cylinder numbers) 1-3-4-2 firing order is the 'rule of nine'. For example, if the engine is turned until valve number one is seen to be fully open, the gap on valve number eight can be checked/adjusted, for this valve will be fully closed; similarly, if, say, valve number six was fully open, you would check/adjust valve number three. In other words, the second valve number, added to the first, will always make nine.

(On a six-cylinder engine this is the 'rule of thirteen'.)

There are variations on the theme and, indeed, with experience your observation will tell you that as you turn the engine to fully open a specific valve, another will also fully open at the same time, allowing you to check/adjust two closed valves at a time, and thus cutting down on the number of rotations of the engine that need to be made.

You can jot down the valve numbers, establish a logical engine rotation sequence, and tick the valve numbers off your list as you check/adjust them.

If you are uncertain as to whether any short cut applies to your engine, then the 'long way round' at least has the virtue of being foolproof. Here, you turn the engine until, say, number one valve is fully open, then turn the engine one complete revolution - number one valve will now be fully closed and there should be a gap between it and its rocker arm.

17G. To check the adjustment, insert the correct value feeler gauge (e.g., 0.012 in.) into the gap: it should - but only just - be possible to slide the feeler back and forth - it's a good idea to bear down on the other (adjuster) end of the rocker arm at the same time, to ensure all play is taken up and the gap is at its widest.

If you can't get the feeler in, the gap is too small; if the feeler flops about the gap is too large.

To adjust the gap, slacken off the adjuster locknut with a ring spanner, then use a screwdriver (usually) to turn the adjuster screw up or down as required.

17H. This diagram shows schematically how the valve clearance in an overhead-camshaft engine, refered to earlier, works in practice.

INSIDE INFORMATION: Hydraulic tappets are adjusted automatically on these engines (see your handbook) and no adjustment is necessary or possible. (Illustration, courtesy V L Churchill/LDV Limited)

□ Job 18. Air cleaner element.

If the air filter element becomes congested with soot, dust and oil, the airflow into the engine will be restricted, and this will cause reduced performance and increased fuel consumption.

The air filter should certainly be checked annually, and probably is best renewed at this time too, although some schedules suggest you check it now and renew it at around 24,000 miles. High mileage engines with heavy oil consumption are likely to make their filter go very sooty in a short space of time.

18A. Generally the element will be found beneath a lid on the air cleaner body, the lid held perhaps by a central nut, perhaps by a ring of wire clips (as on this Citroen XUD engine), perhaps by the nut and clips, perhaps by three or four nuts or bolts. You will invariably find out just by looking!

18B. On this Vauxhall engine, the filter housing is half formed by a cast metal plenum chamber.

18C. INSIDE INFORMATION: Air ducting is so often taken for granted throughout the life of the vehicle, and as a result is never checked. However a blocked air duct will have the same dramatic effect on engine performance as a blocked filter. All air ducting must be examined at regular intervals for kinking, collapse or blockage, and at the same time checked that it is correctly located and secure. (Illustration, courtesy V L Churchill/LDV Limited)

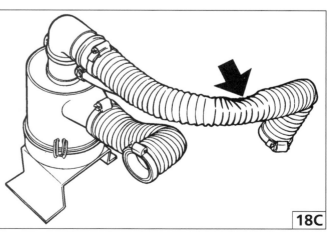

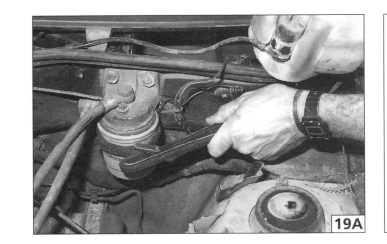

19A

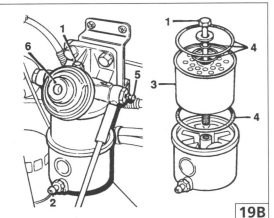

19B

☐ Job 19. Renew fuel filter.

Also, see *FACT FILE: FUEL FILTER* before *Job 7.*

Impurities carried in the diesel fuel cannot be tolerated as they can caused internal damage to the injection pump and injectors. An extremely fine fuel filter is therefore connected in line with the fuel supply to the injection pump, capable of restricting the progress of even microscopic air-borne impurities such as dust particles.

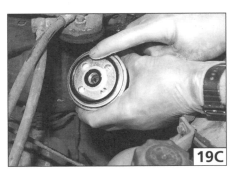

19C

If the filter becomes clogged with impurities fuel flow can become reduced, so impairing engine performance. Filter replacement is therefore vital at specific intervals - often at 12,000 miles/12 months, though sometimes as high as every 18,000 miles/18 months. We always favour the more frequent interval for a fuel filter change.

Like the conventional cartridge-type engine oil filter, most diesel fuel filters are disposable and screw hand-tight to a filter head, sealing by means of a rubber ring. Some (namely certain Lucas filters) are of a two-part design in which the filter base/water bowl or canister has a threaded centre tube running up through the filter element and bolting through the filter head.

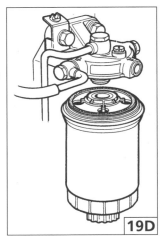

Quite a few filter heads carry a hand-operated priming pump which allows the filter to be primed with fuel, and also facilitates the purging (bleeding) of air from the fuel system.

SAFETY FIRST!
Whenever you are dealing with diesel fuel, it's essential to protect your hands by wearing plastic gloves.

19D

19A. If the filter is of the one-piece cartridge type, it is removed by unscrewing by hand, or if too tight, with the aid of an oil filter wrench.

19B. Alternatively, removal is a question of undoing the centre nut/bolt (1). Note the sealing rings (4); ensure that the small one is in the filter box when you buy the replacement cartridge (3)! (Illustration, courtesy V L Churchill/LDV Limited)

19C. The rubber sealing ring at the top of the filter element or cartridge should be smeared with diesel fuel prior to fitting so that it doesn't grip and prevent proper tightening.

19E

19D. Note that several designs of filter, such as the Delco type fitted to many GM models, and the Bosch type shown here have a small diameter inner seal as well as a larger outer one, plus a plastic retainer to secure it prior to fitting the filter. (Illustration, courtesy V L Churchill/LDV Limited)

INSIDE INFORMATION: On some fuel systems, particularly those which are not provided with a hand priming pump, it is a good idea to fill the fuel filter assembly with diesel fuel before reassembling it. This will reduce the amount of air-bleeding required, and will make for quicker, easier engine-starting.

19E. The new filter is screwed firmly into place by hand only, or, if of the bolt-retained type as shown here, tightened gently with a spanner.

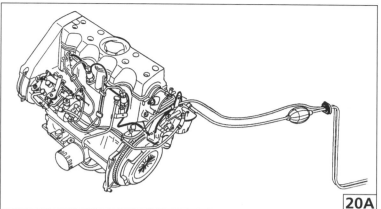

20A.

☐ **Job 20. Fuel system air bleeding.**

When a fuel filter has been replaced, it is important to bleed air from the filter assembly before starting the engine. Air in the fuel system gives rise to difficult starting and erratic engine running, though if present in tiny bubbles only it eventually bleeds itself out via the injection pump return line as the engine runs.

20A. If the car has a priming bulb separate to the fuel pump, you prime the system by pumping the bulb, rather like the doctor taking your blood pressure. (Illustration, courtesy Peugeot)

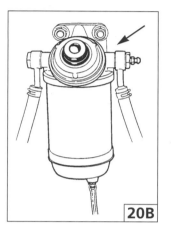

20B.

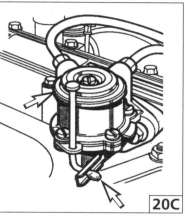

20C.

20B. If the car doesn't have a fuel supply pump external to the fuel injection pump (i.e. most cars) it probably has a manual priming device either on the filter head, or somewhere in the engine bay, in-line with the fuel supply pipe or hose. (Illustration, courtesy Peugeot)

20C. An external fuel supply pump usually has its own priming lever (arrowed). (Illustration, courtesy V L Churchill/LDV Limited)

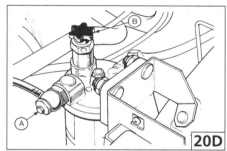

20D.

20D. A primer allows fuel to be drawn from the tank without the engine running, and there is usually an air bleed hole (with a screw plug in it) on the filter head to allow aerated fuel to emerge while priming by hand. (Illustration, courtesy Ford Europe Ltd)

20D. In rare cases where no primer is fitted, bleeding can be achieved by cranking the engine with the starter motor, though to prevent the engine from starting the injector unions should first be slackened very slightly and surrounded with rags to absorb wasted fuel.

20E.

20E. In addition to bleeding air from the fuel filter vent, bleeding can also be carried out from the injection pump fuel inlet union and the injector unions if air has been absorbed that far into the system. As with water, air can be seen as bubbles in the fuel, and as soon as bubble-free fuel emerges, the vent screw or union can be retightened.

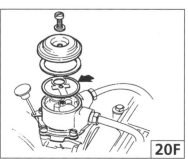

20F.

INSIDE INFORMATION: It is best to hold the primer on the downstroke (or compressed, depending on primer type) when retightening, to prevent air from being drawn back into the system.

20F. Don't forget that certain lift pumps such as the AC and the Super Par do have a servicing requirement. At the specified intervals, remove the top cover, remove the wire mesh filter wash it clean and refit.

FACT FILE: COMPUTER PROTECTION.

Many vehicles depend on a constant power supply from the battery and you can find yourself in all sorts of trouble if you simply disconnect the battery on those vehicles. You might find that the car alarm will go off, you could find that the engine management system (applicable only to a very few state-of-the-art diesels) forgets all it has ever "learned" and the car will feel very strange to drive until it has re-programmed itself, and you could find that your radio refuses to operate again unless you key in the correct code. And if you've bought the car second-hand and don't know the code, you would have to send the set back to the manufacturer for re-programming. So, you must ensure that the vehicle has a constant power supply even though the battery is removed. To do so, you will need a separate 12 volt battery supply. You *could* put a self tapping screw into the positive lead near the battery terminal before disconnecting it, and put a positive connection to your other battery via this screw. But you would have to be EXTREMELY CAREFUL to wrap insulation tape around the connection so that no short is caused. The negative terminal on the other battery would also have to be connected to the car's bodywork.

A better way is to use something like the Sykes-Pickavant

Computer Saver shown here. Clip the cables to your spare battery and plug it into your cigarette lighter socket. (You may have to turn the ignition switch to the "Auxiliary" setting to allow the cigarette lighter socket to function.)

You have to hold in the red button on the Computer Saver while inserting it into the lighter socket, and if two green lights still show after the button is released, you have a good connection and your battery can now be disconnected and removed.

Be sure not to turn on any of the car's equipment while the auxiliary battery is connected.

□ Job 21. Battery terminals.

Provided the battery is kept clean and dry and is not topped-up over generously, its terminals should also remain clean and sound unless a generator fault is causing it to be over-charged, with consequent heavy 'gassing' from the cells.

SAFETY FIRST!
i) The gas given off by a battery is highly explosive. Never smoke, use a naked flame or allow a spark to occur in the battery compartment. Never disconnect the battery (it can cause sparking) with the battery caps removed.
ii) Batteries contain sulphuric acid. If the acid comes into contact with the skin or eyes, wash immediately with copious amounts of cold water and seek medical advice.
iii) Do not check the battery levels within half an hour of the battery being charged with a separate battery charger because the addition of fresh water could then cause the highly acid and corrosive electrolyte to flood out of the battery.

Many vehicles were fitted by the manufacturer with a 'sealed for life' battery, but it is possible that yours may have a 'normal' item, fitted by a subsequent owner. If yours is the former type, then no maintenance is required. This section relates to the common type of replacement battery.

21A. Generally speaking, it is electrolyte spillage or excess vapour which leads to the 'fungal growth' noted on the terminals of neglected batteries as seen here. It is a condition which, as well as the highly corrosive effect on nearby metals, such as the battery clamp and the battery tray, also causes poor electrical contact. In the extreme, the starter may fail to operate, or all electrics may apparently fail.

21A

21B. If you have inherited a second-hand vehicle suffering from this problem, simply pouring hot water over the terminals and any other affected parts, such as the battery strap or clamp, and the battery tray, will usually prove remarkably effective. Take care that you don't pour the hot water into the battery cells or onto nearby vulnerable components and ensure all of the corrosive 'fuzz' is washed away from the car's bodywork.

21C. If necessary, the hot water treatment can be followed by the use of a wire brush and/or emery cloth to bring the battery lead connectors and the battery terminal posts back to clean and bright condition. A slim knife-blade or half-round file can be useful to clean inside the lead connectors.

Be very careful to guard against 'short circuits' when working on battery terminals. The gas ensuing from the cells, particularly when the battery is being charged, is extremely explosive and ignition by a careless spark can cause a truly horrific battery explosion.

(A typical cause of accidental short-circuit is the bristles of a wire brush touching a battery terminal and a battery strap at the same time, or, similarly, a spanner being used to tighten a terminal nut also touching this strap or the car bodywork.)

21D. Once all connections are clean and dry, a smear of petroleum jelly, a proprietary battery jelly or copper-impregnated grease, will guard against further corrosion and help to maintain good electrical contact. Badly affected metals should be treated with a rust killer and re-painted. Smear lightly inside; more heavily over all the outside of the terminals.

21E. Finally, terminal connections should be tight but not 'murdered'.

24,000-mile/24-month Service

Carry out the 12,000-mile/12-month service operations, plus the following:

The Engine Bay

☐ **Job 22. Coolant renewal.**

> *SAFETY FIRST!*
> *i) If you remove the pressure cap when the engine is hot, the release of pressure can cause the water in the cooling system to boil and spurt several feet in the air with the risk of severe scalding.*
>
> *ii) Take precautions to prevent anti-freeze being swallowed or coming in contact with the skin or eyes and keep it away from children. If this should happen, rinse immediately with plenty of water. Seek immediate medical help if necessary.*

Once your cooling system contains anti-freeze, your diesel car should be protected from the dangers of freezing up. However, there is the danger that topping-up the coolant with plain water will weaken the mixture. On top of that, the beneficial 'all the year round' corrosion inhibitors built into the mixture will also be weakened. Therefore, the anti-freeze/coolant mixture should be changed at least every two years, while many motorists are happier to still regard the job as an annual pre-winter precaution.

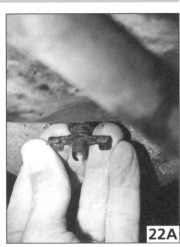

And, of course, if you have recently bought a second-hand vehicle it would anyway be wise to start afresh, although you could take it to a garage and ask them to check the strength of the anti-freeze for you (a simple 'dip' test) or indeed, you could buy yourself a suitable hydrometer from your local accessory shop.

If you decide to renew the mixture in your car's cooling system, first consult your handbook to determine the cooling system's capacity - alongside that data the handbook might even recommend (as a percentage) what the anti-freeze content should be.

22A

The biggest problem now is likely to be the draining out of old coolant.

22A. On older cars, drain taps on both the radiator and the engine cylinder block are easily identifiable and easily found.

22B. Other older cars might have some sort of drain tap or a plug in the bottom of the radiator, but you might be hard put to find anything in the way of a drain plug, let alone an identifiable tap, on the engine block.

22C. For the vast majority of cars, the only way to let coolant out will be to temporarily disconnect the bottom hose from the radiator.

INSIDE INFORMATION: Every few years, you may want to flush the system. Disconnect top and bottom hoses from the radiator. Insert the garden hose, first into the bottom of the radiator, then into the end of the bottom hose (heater taps open), stuffing the gap with a piece of rag. Then flush again from above. You'll be surprised by how much muck you'll shift and your heater could even work better afterwards!

22B

It used to be widely held that anti-freeze causes leaks. Basically, it doesn't, it finds them: having a lower surface tension than ordinary water, it can flow out through gaps tight enough normally to contain the plain water - hence the need to check hoses and clips before adding anti-freeze.

Anyway, when you do drain, by whatever means, first remove the filler cap (from radiator or expansion tank, as appropriate) to aid the flow - and, of course, never attempt the job until the coolant has cooled sufficiently for there to be no danger of scalding yourself. Heater controls should be set to 'Hot', incidentally, so that this part of the system is also open to coolant flow.

22C

22D. When you are happy that the system is ready to receive its fresh anti-freeze, it is usually more sensible (as already mentioned) to first add the required quantity of neat anti-freeze, then top up with clean water as required to bring the coolant content to its correct level. If the radiator has a cap, pour the anti-freeze into the radiator...

22E. ...if the radiator doesn't have a cap, pour the anti-freeze into the expansion tank.

Should you be unlucky enough to find that a little less than your proposed amount of neat anti-freeze fills the system to capacity, you might be able to console yourself (bearing in mind our average winter temperatures) that a 25 per cent solution should protect down to minus 13 degrees Centigrade - it's very unlikely that you won't have got enough in to give yourself peace of mind - but do remember the higher anti-corrosion requirements of some engines. Check your handbook.

INSIDE INFORMATION: Whatever you do, never mix a coolant solution any stronger than 50% anti-freeze/water, as - remarkably - the anti-freezing properties actually diminish above 60% concentration.

Pour slowly and carefully, so the system gradually fills without danger of air-locks building up anywhere. When the level appears to be correct, start the engine and let it run for a few minutes with the filler cap off. Note that as the thermostat starts to open (evidenced by the radiator top hose warming up) any minor air-locks may disperse, causing the level to drop back a bit - when feeling the top hose, beware the fan or other moving parts.

Top up as required, stop the engine, and if all appears to be well, replace the cap.

In the unlikely event of a serious air-lock, however, this could cause the level to rise and threaten an overflow, so be ready to replace the cap and switch off immediately.

Such an air-lock is often betrayed by one or more of the heater hoses still feeling cold (or at best only luke warm) even when the engine has been run for a while - the heater hoses should warm up and, indeed, feel quite hot some while before that top hose will start to feel warm. Again, take all due precautions when feeling the hoses.

INSIDE INFORMATION: A crude, but effective way of dispersing this air-lock, having now stopped the engine, is to first loosen the heater return hose at its junction within the engine compartment - this is the one that probably leads to the radiator bottom hose stub or an adjacent stub at the water pump.

22F. On some cars it might be easier to detach it at the heater unit connection, but make sure it is the return hose, and not the feed hose from the engine.

22G. Now run the engine and retighten the hoses almost immediately in the hope that air has been expelled. Try raising the front of the car on ramps and running the engine for a minute or two, as another means of dislodging air. Also, try revving the engine several times.

Top up the coolant level now as required, and check it after every run for the next few days to ensure that it has settled down, and that no leaks are occurring anywhere.

Retain a water/anti-freeze mix of the recommended concentration for any topping-up required in the future.

IMPORTANT NOTE: Be sure not to run a diesel engine without water in the block or head - and in particular around the injectors - or permanent damage will be caused. Have a specialist carry out this work if necessary.

☐ Job 23. Checking/adjusting idle/fast idle speed

The only injection pump adjustments which are required on some service schedules are the idle (tickover) and fast-idle speed adjustments.

23A. The idle speed is set by the idle speed screw on the injection pump (situated on top of distributor pumps)... (Illustration, courtesy Ford Europe Ltd)

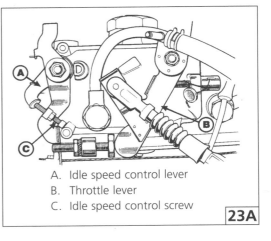

23B. ...and usually towards the end of the governor housing on in-line pumps (arrowed). This screw is simply a stop screw against which the throttle linkage rests. It is very important not to confuse this screw with the maximum speed screw which is invariably situated nearby, as this latter acts as a limit stop to the accelerator lever (or the internal control rod in the case of the in-line pump). The maximum speed screw determines the maximum quantity of fuel that can be injected into the engine, thus preventing it from over-speeding and damaging itself. (Illustration, courtesy V.A.G.)

SPECIALIST SERVICE: THE MAXIMUM SPEED SCREW SHOULD NEVER BE TOUCHED! If you're not sure which is which, have your local specialist either explain to you - or carry out the job for you. Idle adjustment (for which you ideally need a suitable tachometer) is a simple enough task that should be carried out only when the engine is at operating temperature. The idle adjustment (like most adjustments on a diesel) is one of those things that rarely ever needs doing, but if you think the engine is ticking over too fast or too slow, by all means set it right. In most cases it is simply a one-screw adjustment - clockwise to raise the speed and anti-clockwise to lower it - but a locknut may be fitted to the screw, so loosen this first and be sure to fully retighten it when you're done.

A. Idle speed control lever
B. Throttle lever
C. Idle speed control screw

23A

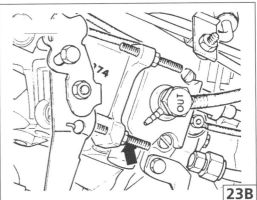

23B

INSIDE INFORMATION: If you do not have either a diesel test-tachometer or a facia-mounted tachometer on your car, yet you feel the idle speed is obviously too high or too low, there is no reason why you shouldn't adjust the speed by ear until it is at a satisfactory level. On early diesels this was best accomplished by increasing the idle speed until the dashboard was no longer a blur!

On some pumps the idle adjustment is more complicated than this, and may be tied in with adjustments of the cold-start fast idle, and something called anti-stall or deceleration adjustment. Specific adjustment details must be sought for such arrangements or **SPECIALIST SERVICE.**

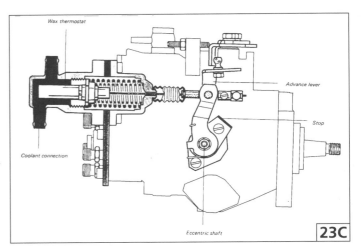

Wax thermostat

Advance lever

Stop

Coolant connection

Eccentric shaft

23C

23C. Some diesels have a wax capsule which contracts or expands depending upon engine temperature, pulling a cable which lifts the injection pump throttle lever off its stop to give a fast idle when cold. Adjustment of this linkage may be independent or tied in with normal idle adjustment, so specific details are needed or **SPECIALIST SERVICE.** (Illustration, courtesy V.A.G.)

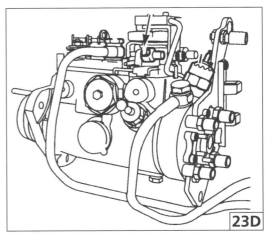

23D. Some injection pumps - particularly of Lucas manufacture, feature an anti-stall function - an adjustable stop in the top of the pump (arrowed), which keeps fuel delivery high enough to prevent stalling when the throttle is suddenly released. Adjustment of this stop is usually tied in with normal idle adjustment, so specific adjustment details should be referred to or **SPECIALIST SERVICE.** (Illustration, courtesy Peugeot)

23D

36,000-mile/36-month Service

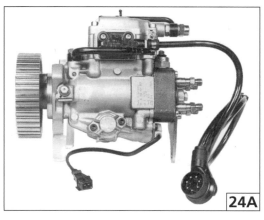

24A

Carry out the 24,000-mile/24-month service operations, plus the following:

The Engine Bay

☐ **Job 24. Checking/adjusting injection timing**

OPTIONAL OR SPECIALIST SERVICE

24A. Correct injection timing is as important to smooth, powerful, efficient diesel engine operation as accurate ignition timing is to the proper functioning of the petrol engine. Unlike the latter, which is notorious for going off-tune - mostly thanks to ignition points gap variation, distributor wear and spark plug fouling - the diesel tends to remain consistent where timing is concerned, thus rarely requiring timing checks or adjustment. Rarely, if ever, will you find a diesel car service schedule mentioning a check on injection timing. Illustrated is actually an electronically controlled pump - a **SPECIALIST SERVICE** item.

Yet there invariably comes the time when an injection timing check is called for, due maybe to the requirement for fault diagnosis or the unavoidable disturbance of the injection pump position because of some other engine repair work - namely a change of timing belt.

Injection timing determines the correct relationship between the position of the pump injection element and that of the engine's crankshaft. This relationship is established on one cylinder (usually No.1) just as it is with the rotor arm-to-crankshaft relationship on petrol engines. Once set, it is correct for all cylinders.

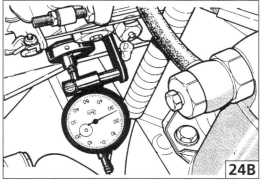

24B

With the piston of cylinder No.1 positioned at a specified number of crankshaft degrees (or millimetres of travel) before Top Dead Centre on the firing stroke, the injection pump plunger must be at a particular point on its pumping stroke for timing to be correct.

24B. On the common (distributor type) injection pumps this position is usually measured with a dial gauge. This is a Lucas pump ... (Illustration, courtesy Ford Europe Ltd)

24C. ... and this is a Bosch pump. On the much less common (in-line type) pump found on some V.A.G., Mercedes-Benz cars and some Japanese off-road cars), the determination of plunger position generally involves a different procedure - which we'll discuss shortly. (Illustration, courtesy V.A.G.)

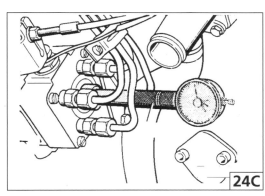

24C

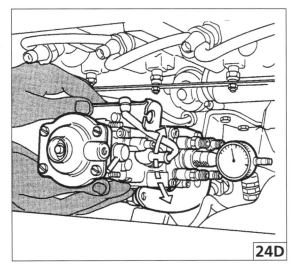

24D

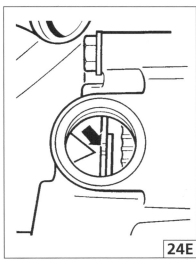

24E

24D. The position of the high-pressure plunger is varied either by turning the injection pump body or by adjusting a two-part pump drive hub (in cases where the injection pump is fixed rigidly to the engine at both ends). (Illustration, courtesy Fiat)

24E. Crankshaft position is most commonly determined by Top Dead Centre and/or injection point (i.e. timing) marks provided at the flywheel ... (Illustration, courtesy V.A.G.)

24F. ... or the crankshaft pulley or flywheel. (Illustration, courtesy V.A.G.)

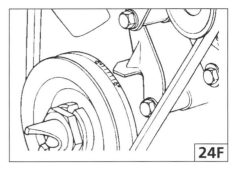

24G. In some cases, though (Ford Fiesta/Escort/Sierra, and Peugeots/Citroens with the popular XUD engine, for instance) the crankshaft is set in the relevant position by means of a locating pin inserted through the crankcase. (Illustration, courtesy Ford Europe Ltd)

24H. On some other engines, such as the 2.3-litre Peugeot unit (also used in the Ford Sierra), the correct crankshaft position is determined by dropping the exhaust valve of No.1 cylinder onto the piston, then measuring valve travel with a dial gauge as the piston moves in the cylinder. This technique may sound daunting - and it is - though fortunately it's the exception rather than the rule! (Illustration, courtesy Ford Europe Ltd)

24I. It is not uncommon for the injection pump mounting flange to carry an alignment mark corresponding to another on its mounting bracket. A pump which has been moved can be realigned to these marks to regain the correct timing, provided the crankshaft has not been turned in the interim. If no such marks exist, a fine line can be scribed between pump flange and mounting as an installation reference before the pump is moved. (Illustration, courtesy Ford Europe Ltd)

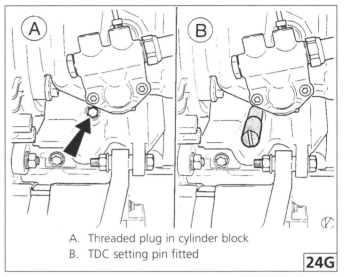

A. Threaded plug in cylinder block
B. TDC setting pin fitted

Injection timing can be carried out statically or dynamically, although the latter method necessitates fairly expensive equipment and is not always advocated by the manufacturer, despite the fact that it is more accurate. As there are literally scores of different injection timing procedures in existence, some apparently very pernickety, others surprisingly straightforward. We'll outline the most common techniques used.

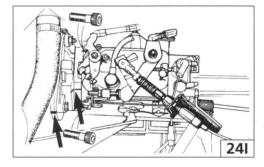

STATIC TIMING

24J. This is the method most likely to be employed by the DIYer, and in most cases it requires investment in a dial gauge and holder (NOT an inexpensive item to purchase!)...

24K. ...although the fortunate owner of an in-line injection pump will probably get away with the purchase of a spill-timing tool instead, and will get a lot more change back!

The Bosch (and Japanese pattern) VE distributor type injection pump and most Lucas distributor pumps (those cover over 90% of diesel cars!) have a plate or plug to allow access for fitting a dial gauge or a timing pin to determine the exact position of the high pressure plunger.

24L. On the Bosch VE pump the plug is at the rear in between the injector

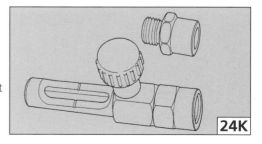

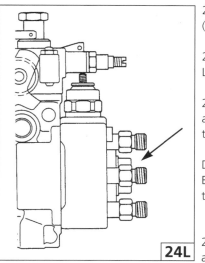

24L. On the Bosch VE pump the plug is at the rear in between the injector pipes... (Illustration, courtesy V L Churchill/LDV Limited)

24M. ...while on Lucas pumps the plate is fitted on the pump flank. (Illustration, courtesy Lucas)

24N. Some early Lucas DPA type pumps have an internal mark (on the pump rotor) that aligns with another on the housing, and where this is the case the pump plunger is positioned purely by alignment of these marks with no recourse to special equipment.

Dial gauges, holders and timing pins are manufactured by injection system manufacturers Bosch and Lucas, as well as many tool and equipment manufacturers who supply the trade - see *Chapter 9, Tools & Equipment.*

24O. With in-line injection pumps, determination of the injection point may be by alignment markings (arrowed) on pump flange and mounting bracket...

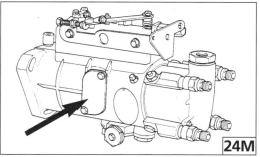

24P. ...by the insertion of a timing pin through the pump drive gear and into the bracket.

24Q. Then, a dial gauge is used to measure the lift of the pump plunger, or it can be measured by the spill-timing method illustrated here. Many in-line pumps are timed using this last simple technique in which the injection point is found by removing the injector pipe and delivery valve (this sits in the pump-to-pipe union) from the relevant cylinder and substituting it with a short, open pipe in the shape of a 'U'. The injection point can be determined by keeping the injection pump pressurised with fuel and turning it in the specified direction of rotation until fuel emerges from the spill pipe, then backing off very slowly to the precise point at which fuel stops dripping - this cut-off point is also the injection timing point. (Illustration, courtesy V L Churchill/LDV Limited)

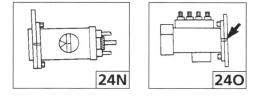
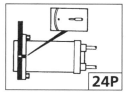
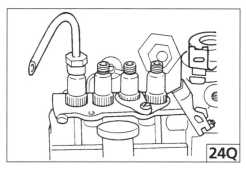

DYNAMIC INJECTION TIMING

This is carried out with the use of a diesel timing light in conjunction with the engine's timing or Top Dead Centre reference mark. Dynamic timing provides a more realistic indication of injection point because it does so under operating conditions. However, not all engine manufacturers consider dynamic timing a necessity, so the relevant timing figures aren't always available.

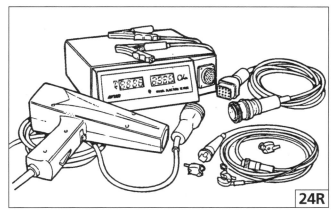

24R. The diesel timing light receives its triggering pulse either from the relevant injector pipe or from a flywheel sensor. Certain diagnostic units can accept an input signal from an injection point sensor (fitted only to a few in-line injection pumps - e.g. Mercedes-Benz) but it is more common to take a pulse from the injection pipe by means of a pressure-sensitive clamp that fits around it. (Illustration, courtesy V L Churchill/LDV Limited)

The timing is checked by shining the timing light at the timing mark on the flywheel or crankshaft pulley, and it is adjusted by turning the injection pump body until the fixed and moving marks appear to coincide - just as dynamic ignition timing is achieved on the petrol engine.

Where there is no timing mark, but a Top Dead Centre reference instead, a timing light with an advance function is needed. This allows the user to advance the flash of light by as many degrees as the quoted timing value of his engine - say, 10 degrees Before Top Dead Centre - by putting it out of phase with the triggering pulse from the engine. This enables the TDC mark to be used as the true timing mark.

Electronic diagnostic units are available to the trade (unfortunately not cheaply) which embody dynamic timing, timing advance checking, and engine rpm checking in one device. With an advance-type diesel timing light costing very dearly, and dynamic timing equipment generally being expensive, static timing is the natural choice for DIYers. However, at least a couple of companies market an electronic adaptor unit which allows a run-of-the-mill petrol timing light to be used with either a clamp-on injector pipe pick-up or an optical pick-up that fits into the engine in place of a glow plug - see *Chapter 9, Tools & Equipment.*

As dynamic timing checks have to be made at specific engine speeds, there must be a means of accurately gauging engine speed. Of course a conventional test-tachometer cannot be used on a diesel engine as there is no ignition system from which to operate it. Some diesel cars are equipped as standard with a facia-mounted tachometer but it is not wise to refer to these for the purposes of engine adjustment as they aren't sufficiently accurate.

Some diesels have a flywheel TDC sensor which can be connected to suitable electronic test equipment (again, equipment that is likely to be beyond the average DIY budget) via a diagnostic socket in the engine compartment in order to provide a speed reading.

The keen DIYer's best solution, though, is to obtain a hand-held optical tachometer - a device which registers the number of times that a white mark made on the crankshaft pulley or flywheel passes it - see *Chapter 9, Tools & Equipment.*

24S. For the real enthusiast, though, (and you have to do a lot of diesel tweaking to justify this) the injection-pulse sensitive tachometer (which uses the same clamp-on pick-up as the timing light mentioned previously) is available either as a separate, single-purpose unit, or as an included function of a multi-function electronic test unit. (Illustration, courtesy V L Churchill/LDV Limited)

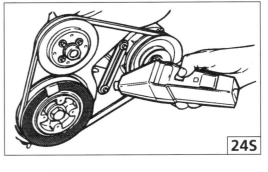

24S

☐ Job 25. Timing Belt Inspection/Renewal

OPTIONAL

25A. Most modern diesel car engines, such as this Golf diesel unit, have their camshaft and injection pump driven by a toothed rubber belt, the timing belt (arrowed). Some, though, use a steel chain-drive... and others, a gear-wheel drive.

It is particularly important to renew the timing belt at manufacturers' specified intervals as a belt which breaks or strips its teeth while the engine is running will invariably cause internal engine damage. This damage, which is the result of the pistons and the valves in the cylinder head hitting each other, can also occur on petrol engines, although not as often, as petrol engines have a far greater clearance between valves and piston when the piston is at the top of its travel.

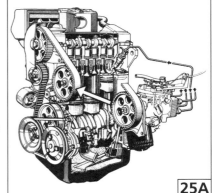

25A

Depending on the accessibility of the timing belt (many are hidden behind complicated belt cover arrangements) it may be a good idea to inspect the belt condition at the time of each major service (typically once a year).

If a belt is subjected to excessive road dirt (due to a missing cover) or, more importantly, to spilled diesel fuel or leaking engine oil, it will rapidly deteriorate, this deterioration being evidenced by splitting, cracking, missing teeth, or flabbiness.

Signs of any of these should be addressed by a belt replacement, and if oil leakage is the cause, this must be rectified before the new belt is fitted.

25B. Some manufacturers' timing belt replacement intervals are as generous as 80,000 miles, while others are as cautious as 35,000 miles. Our general recommendation is that timing belts should be renewed at around the latter mileage. When you see what a tortuous path a typical belt has to follow, you can appreciate why. And a broken belt will lead to a wrecked engine! (Illustration, courtesy Peugeot)

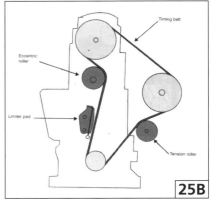

25B

48,000-mile/48-month Service

Carry out the 36,000-mile/36-month service operations, plus the following:

The Engine Bay

☐ **Job 26. Glow plug renewal**

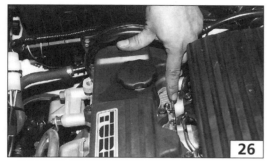

26. While you'll rarely see glow renewal intervals specified in service schedules, a good general rule is that by 30,000 miles the plugs are past their prime, so we, along with Lucas, recommend their renewal at this mileage. And that means ALL plugs, not just those that look ropey!

Glow plug removal and installation is detailed in *Job 16* under the 12,000-mile service operations, and pre-heating system checks are discussed in *Chapter 6, Fault Finding.*

Job 27. Injector maintenance

SAFETY FIRST!
Whenever you are dealing with diesel fuel, it's essential to protect your hands by wearing plastic gloves.

NEVER work on the fuel pipework when the engine is running. ALWAYS place a rag over the union when it is undone. Very high pressure is retained for a considerable time after the engine is last used.

OPTIONAL AND SPECIALIST SERVICE

The correct operation of the injectors is vital to the smooth and efficient running of the diesel engine. Anything that interferes with the injector spray pattern, opening pressure, nozzle leak-tightness or injector back-leakage (controlled leakage of fuel around the nozzle needle and out to the fuel return lines) is ultimately to the detriment of proper combustion. Engine symptoms of poor injector performance include knocking noises, overheating, misfiring, black exhaust smoke (with the attendant increase in fuel consumption) and loss of power and smoothness.

Injectors tend to provide long, trouble-free service. There does, however, come a time in the life of an injector when its performance is reduced by deposits of carbon and gum in and around the nozzle. Although it would be unwise to state that this occurs at a particular mileage, most injectors 'keep a low profile' for 70,000 miles, and some keep on performing well beyond this mileage. If you can stretch to the cost and/or time, we feel it's worth checking on injector performance at or around the 48,000-mile service.

Few, if any, service schedules include injector maintenance, and although it is probably sufficient to leave well alone until you have reason to suspect that there may be injector problems, it doesn't cost a great deal to whip the injectors out at service time, clean the nozzles and maybe also check the spray characteristics. Peugeot/Citroen XUD engine specialists tell us that an injector overhaul at 70,000 miles tends to prove beneficial to smooth running and fuel economy. Whether you personally wish to get involved with injector checking and overhaul is entirely dependent on your level of enthusiasm, the equipment available, and your devotion to preventative maintenance!

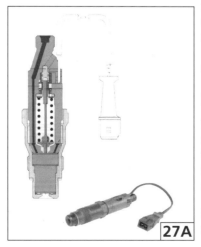

REMOVING INJECTORS

When removing injectors it's worth remembering the following points. Firstly, before undoing the injector unions, make sure that they are free of harmful dirt which could enter the fuel lines. Unscrew the fuel pipe union at the injector. If the union is of the common flare-nut type (as used on brake pipes) undo it using a flare nut wrench - this is a ring spanner with a split in it to accommodate the injector pipe. (See Job 20D.)

27A. Also remember that some might have electrical connections which need to be disconnected. This Lucas injector has an inductive sensor which corrects the timing when necessary and is usually fitted only to number 4 cylinder. (Illustration, courtesy Peugeot)

27B. Next, loosen the union at the injection pump end of the pipe. Disconnect the fuel-return unions at the injector and move the return pipes away.

27C. Threaded injectors (by far the most common type on cars) should be unscrewed either with a proper injector removal socket, or with a conventional socket that is suitably long; the standard size is 27 mm AF (across flats).

IMPORTANT NOTE: If the injector is of the clamp or flange-mounted type, however, secured by more than one nut, loosen the nuts evenly so as not to distort the injector nozzle. With the clamp removed, try to pull the injector from the cylinder head. If it is tight, you will need to use an injector puller - see *Chapter 9, Tools & Equipment.*

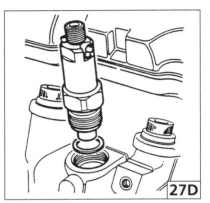

27D. Most injector bodies (A) are sealed against the cylinder head with a copper ring (B) that seals the shoulder of the injector nozzle, though the nozzle itself sometimes sits on a dished washer at the bottom of the injector bore. All seals should be renewed every time the injector is removed. (Illustration, courtesy Lucas)

The injector bore should always be cleaned carefully before an injector is refitted as any dirt can cause leakage of cylinder compression, as can the use of an old, flattened seal, and leakage can lead to expensive erosion of the injector by hot, escaping gases. When refitting screw-type injectors, observe the specified tightening torque, or tighten firmly without using excessive force or too long a drive lever on the socket.

27E. Once the injector is out, take a look at the nozzle end and see if it is caked with carbon - even a tiny amount of the stuff can upset that critical fuel spray pattern. (Illustration, courtesy V L Churchill/LDV Limited)

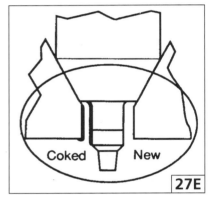

27F. There's certainly no harm in brushing around the pintle of an indirect-injection type injector (the pintle is the central protruding tip) or the holes of an indirect-injection engine with a brass brush (a suede shoe brush is ideal for this). You can even poke a strand of brass wire into the holes to ensure they're clear. But DON'T ever be tempted to use a steel brush or strands on any part of the injector.

SPECIALIST SERVICE. Professional injector cleaning kits are available. These kits include useful probes and scrapers for internally de-gumming nozzles, but we really don't recommend that any DIYer become involved with injector disman-tling. Remember, you don't have to renew an injector if it's misbehaving, you can take it to a Bosch or Lucas agent for re-nozzling and calibration, and this needn't be very expensive.

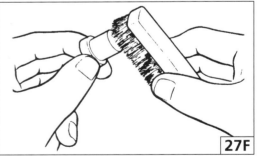

The injector should ideally be removed from the cylinder head and connected to an injector tester for a check on fuel spray pattern.

INJECTOR FUEL SPRAY PATTERN CHECK

27G. An injector tester is essentially a bench-mounted, hand-operated high-pressure pump containing diesel fuel, to which the injector is connected. It is fitted with a pressure gauge so that verification can be made of the pressure at which the nozzle opens (the break pressure). If the break pressure is incorrect engine operation will be impaired. (Illustration, courtesy V L Churchill/LDV Limited)

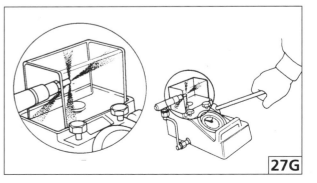

27H.

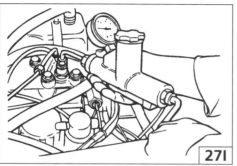

27I.

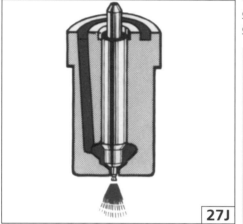

27J.

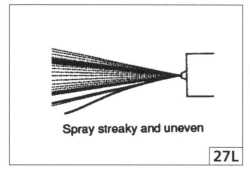

Uniform, well atomised
spray pattern

27K.

27H. And here's a similar set-up in use.

27I. A convenient and affordable alternative to the conventional injector tester, known as the 'Minitest' is available from V L Churchill/LDV Limited. This gadget allows the testing of an injector while it is still fitted to the cylinder head. Of course this doesn't allow sight of the injector spray pattern, but the pattern is by no means the whole story. The Minitest is a hand-held pump which is filled with diesel fuel and connected to the injector. The fuel is pumped up to pressure by means of a crank handle, until the needle of the Minitest's pressure gauge jumps as the nozzle opens. A 'crack' is simultaneously heard. This indicates the break pressure, which can be compared with specification. (Illustration, courtesy V L Churchill/LDV Limited)

In addition to this, the Minitest can check for injector back leakage (by sustaining a fluid pressure below nozzle opening pressure), seat tightness (by the same method), nozzle chatter, and spray condition if you use it with the injector removed (and the proverbial jam jar in tow). This unit gets our five-star rating for practicality and value.

INSIDE INFORMATION: See also **Chapter 1.** By way of a quick and easy check of injector performance, however, it is possible to observe spray pattern with the injector connected to the injection pump, but removed from the engine. This is done by removing the injector, reconnecting it to its high-pressure pipe, slackening the pressure unions of the remaining injectors to prevent the engine from firing, then cranking the engine with the starter motor. Direct the spray into a glass jar held down, onto the cylinder head, so that it can't move as the injector gives a powerful blast of fuel.

SAFETY FIRST!
The spray from an injector can penetrate the skin or eyes and enter the bloodstream - which is, of course, extremely dangerous. For this reason, always point the injector into a glass jar and well away from you or anyone else in the vicinity. Remember that the spray will give a mighty 'kick' which is enough to blast a jar out of your hands. Keep your hands behind the 'business' end of the injector nozzle. Also, wear goggles and protect your hands from diesel fuel by smearing them with barrier cream, and donning protective polythene gloves before exposure. Read Chapter 1, Safety First!, for more essential tips on how to DIY in safety.

27J. A non-uniform, streaky spray pattern indicates that the nozzle is dirty or damaged, while a symmetrical pattern showing no such irregularities is what is ideally required. (Illustration, courtesy Lucas)

27K. Pintle nozzles (the type used on indirect-injection cars) should produce a spray in the shape of a symmetrical, narrow fan, central with the axis of the injector, free of breaks or streaks and free of fuel droplets.

27L. However, if the spray emerges from the nozzle at an angle, in streaks or in a broken shape, a dirty or damaged nozzle is indicated.

Spray streaky and uneven

27L.

27M. The exception to this is the multi-hole nozzle used on direct-injection engines. This type of nozzle produces a number of fine fans of spray. Where this type is being tested, the spray pattern should be checked in the same manner, but the gaps in the spray should be ignored.

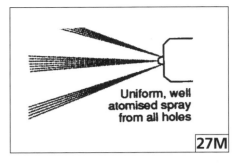

27N. Any lack of symmetry in the spray pattern of a multi-hole injector should be treated with suspicion unless there is clearly a directional bias in the arrangement of the nozzle holes. So take a good, close-up look (but ONLY prior to testing, please!) at the nozzle arrangement so as to get some idea of the type of pattern to expect. (Illustrations, courtesy V L Churchill/LDV Limited)

INSIDE INFORMATION: For the injector assembly to properly atomise fuel, its needle must oscillate at a high frequency during the injection phase. This oscillation can be distinctly heard as a buzzing noise and felt as a vibration through the tester pumping handle. Irregular buzzing or a lack of buzzing can usually be attributed to a poor nozzle condition or a sticking needle. In short, if you are testing an injector that doesn't buzz, you know there's a problem.

The injector nozzle should open sharply at a pre-set pressure (which is nearly always quoted in the car's technical literature) without any prior dribbling of fuel from the nozzle.

SPECIALIST SERVICE. This pressure can only be checked with a proper test device, and is indicated by a flick of the pressure gauge needle as injection occurs. An incorrect opening pressure can be due to dirt on the inside of the nozzle, or a weakened needle spring. Internal cleaning and spring re-calibration can be carried out by a specialist, and this treatment is a cheap, yet satisfactory alternative to purchasing complete new injectors.

INJECTOR NOZZLE LEAKAGE

27O. If you have access to an injector tester, one further check that should be made is for nozzle leakage, though only once the nozzle has been brushed clean. A nozzle leak test value, stipulated as pressure against time, should be available in the manufacturer's technical literature for the car's injection system - consult your main dealer. The injector should be able to maintain a quoted pressure, without leaking fuel from the nozzle when that pressure is sustained for a specified length of time. If there is leakage, pressure will not be maintained, so recourse can again be made to a specialist who should be able to overhaul the injector comparatively inexpensively - all the injectors, in fact, as you can't afford to have imbalanced injector characteristics.

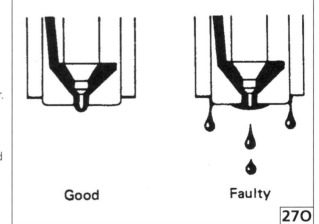

CHAPTER 6 - FAULT FINDING

This chapter aims to help you overcome the main faults that can affect the mobility or efficient operation of your car's diesel engine. It also helps you to overcome the problem that has affected most mechanics - amateur and professional - at one time or another... Blind Spot syndrome!

It goes like this: the car refuses to start one morning. You decide that there must be no fuel getting through. By the time you've actuated the glow plugs a couple of extra times, checked the fuel shut-off solenoid at the injection pump is working okay, and maybe opened up the fuel filter, it's already time for bed. And the next day, the local garage finds that there's air getting into the fuel system, or a wire has dropped off the pre-heating relay! Something like that has happened to most of us!

checks listed here and detailed overleaf, eliminating each check by testing, rather than by hunch, before moving on to the next. Remember that the great majority of failures are caused by air being sucked into the fuel system, or by a battery that is too weak to provide your diesel engine with a decent cranking speed. Follow the sequences that follow, and you'll have better success in finding that fault.

Don't leap to assumptions: if your engine won't start or runs badly, if electrical components fail, follow the logical sequence of

Before carrying out any of the work described in this Chapter please read carefully *Chapter 1, Safety First!*

PART I: DIESEL FAULT FINDER

DIFFICULT TO START (HOT ENGINE)

CAUSE	REMEDY
Fuel starvation	Check fuel/tank/lines/filter
Fuel tank breather obstructed	Clear breather
Air leak into fuel system	Find/stop leak
Fuel contaminated with water or petrol	Empty/refill fuel system
Pump cut-off solenoid inoperative	Repair solenoid/wiring
Incorrect injection timing	Adjust injection timing
Injection pump malfunction	Remove & take to specialist

DIFFICULT TO START (COLD WEATHER)

Fuel waxing	Heat fuel filter
Pre-heat system malfunction	Check circuit & glow plugs
Damaged glow plugs	Renew glow plugs
Fuel starvation	Check fuel/tank/lines/filter
Fuel tank breather obstructed	Clear breather
Air leak into fuel system	Locate and stop
Fuel contaminated with water or petrol	Empty/refill fuel system
Pump cut-off solenoid inoperative	Repair solenoid/wiring
Incorrect injection timing	Adjust injection timing
Injection pump malfunction	Remove & take to specialist

Poor cylinder compression	Head or engine overhaul

ENGINE ONLY STARTS FROM COLD WITH REPEATED PRE-HEAT ACTUATION

Poor glow plug condition	Renew glow plugs
Poor cylinder compression	Head or engine overhaul

ENGINE STARTS THEN STALLS

Fuel contaminated with water or petrol	Empty/refill fuel system
Fuel starvation	Check fuel/tank/lines/filter
Air leak into fuel system	Locate and stop
Insufficient fuel in tank	Fill with fuel
Bad fuel cut-off solenoid connection	Trace and repair

ENGINE IS RELUCTANT TO REV

Fuel starvation	Check fuel/tank/lines/filter
Fuel tank breather obstructed	Clear breather
Incorrect injection timing	Adjust injection timing
Choked air or fuel filter	Renew filter
Air leak into fuel system	Locate and stop
Water or petrol in fuel	Empty/refill fuel system

Worn or contaminated injectors	Clean/overhaul injectors
Incorrect valve timing	Correct valve timing

ERRATIC TICKOVER

CAUSE	REMEDY
Air leak into fuel system	Locate and stop
Incorrect injection timing	Adjust injection timing
Idle speed set too low	Increase idle speed
Worn or contaminated injectors	Clean/overhaul injectors
Water or petrol in fuel	Empty/refill fuel system
Poor cylinder compression	Head or engine overhaul

STALLING OR HESITATION

Fuel starvation	Check fuel/tank/lines/filter
Fuel tank breather obstructed	Clear breather
Incorrect injection timing	Adjust injection timing
Choked fuel filter	Renew fuel filter
Idle speed set to low	Increase idle speed
Fuel waxing (very cold conditions)	Heat fuel filter

MISFIRING

Worn or contaminated injectors	Clean/overhaul injectors
Choked fuel filter	Renew fuel filter
Incorrect injection timing	Adjust injection timing
Air leak into fuel system	Locate and stop
Water or petrol in fuel	Empty/refill fuel system

SURGING ENGINE

Air leak into fuel system	Locate and stop
Excessive crankcase pressure	Clear breather obstructions/overhaul engine
Overfilled sump	Remove some oil

EXCESSIVE FUEL CONSUMPTION

Choked air filter	Renew air filter
Incorrect injection timing	Adjust injection timing
Worn or contaminated injectors	Clean/overhaul injectors
Incorrect valve timing	Correct valve timing
Fuel leak(s)	Locate and stop
Overfuelling injection pump	Seek specialist help
Poor cylinder compression	Head or engine overhaul

BLACK SMOKE

Choked air filter	Renew air filter
Incorrect injection timing	Adjust injection timing
Worn or contaminated injectors	Clean/overhaul injectors
Incorrect valve timing	Correct valve timing
Overfuelling injection pump	Seek specialist help
Engine overcooling	Check coolant thermostat
Poor cylinder compression	Head or engine overhaul

EXCESSIVE KNOCKING/RATTLING

Incorrect injection timing	Adjust injection timing
Worn or contaminated injectors	Clean/overhaul injectors

Poor cylinder compression	Head or engine overhaul

GREY OR WHITE EXHAUST SMOKE

CAUSE	REMEDY
Worn or contaminated injector(s)	Clean/overhaul injectors
Incorrect injection timing	Adjust injection timing
Poor cylinder compression	Head or engine overhaul
Head gasket leak	Renew head gasket
Engine overcooling	Check cooling system
Pre-heating system not effective	Locate/make good pre-heat fault

BLUE SMOKE

Burning engine oil	Head or engine overhaul
May be normal until hot	

ENGINE RELUCTANT TO DECELERATE

Blocked fuel return pipes	Locate/clear blockage
Incorrectly adjusted anti-stall	Set up per specific instructions

ENGINE CANNOT BE SWITCHED OFF

Fuel cut-off solenoid defective	Actuate stop lever on top of pump, or clamp fuel supply hose, then check wiring of/renew shut-off solenoid

ONE OR MORE GLOW PLUGS FAIL FREQUENTLY

Injector malfunction	Remove/check/overhaul injector

ENGINE OIL LEVEL INCREASES (ENGINES WITH TIMING CHAIN OR GEARS)

Injection pump drive shaft seal leaking	Renew drive shaft seal

ENGINE WON'T START (STARTING SYSTEM FAULTS)

As the diesel engine is dependent on energetic cranking if it's to start from cold, the following starting system fault finder is particularly relevant.

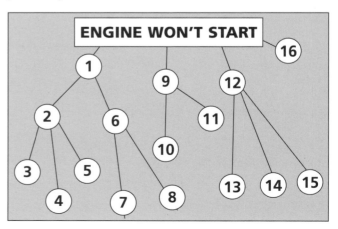

1. Starter motor doesn't turn.

2. Is battery okay?

3. Check battery connections for cleanliness/tightness.

4. Have battery 'drop' test carried out by specialist.

5. Test battery with voltmeter or, preferably, with a hydrometer.

6. Can engine be rotated by hand?

7. If engine cannot be rotated by hand, check for mechanical seizure of power unit, or pinion gear jammed in mesh with flywheel - 'rock' car backwards and forwards until free, or apply spanner to square drive at front end of starter motor.

8. If engine can be rotated by hand, check for loose electrical connections at starter, faulty solenoid, or defective starter motor.

9. Starter motor turns slowly.

10. Battery low on charge or defective - re-charge and have 'drop' test carried out by specialist.

11. Internal fault within starter motor - e.g. worn brushes.

12. Starter motor noisy or harsh.

13. Drive teeth on ring gear or starter pinion worn/broken.

14. Main drive spring broken.

15. Starter motor securing bolts loose.

16. Starter motor turns engine but car will not start. See 'Ignition System' box.

FUEL GAUGE PROBLEMS
17. Gauge reads 'empty' - check for fuel in tank.

18. If fuel is present in tank, check for earthing and wiring from tank to gauge. and for wiring disconnections.

19. Gauge permanently reads 'full', regardless of tank

contents. Check wiring and connections as in 18.

20. If wiring and connections all okay, sender unit/fuel gauge defective.

21. With wiring disconnected, check for continuity between fuel gauge terminals. Do NOT test gauge by short-circuiting to earth. Replace unit if faulty.

22. If gauge is okay, disconnect wiring from tank sender unit and check for continuity between terminal and case. Replace unit if faulty.

PART II FUEL SYSTEM CHECKS

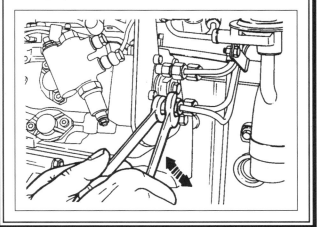

FACT FILE: FUEL INJECTION PIPES

When disconnecting/tightening high pressure pipes, from the pump to the injectors, always prevent the union from turning (arrowed) while turning the nut. Failure to do so might cause breakage. (Illustration, courtesy V L Churchill/LDV Limited)

SAFETY FIRST!
NEVER work on the fuel pipework when the engine is running. ALWAYS place a rag over the union when it is undone. Very high pressure is retained for a considerable time after the engine is last used.

If you've read the diesel fault finder previously, you'll be aware that the majority of potential diesel engine problems are related to the fuel system. Involuntary acceleration, poor starting, hesitation, misfiring, smoke, engine rattling, excessive fuel consumption, poor performance, stalling, vibration... all these and more can result from fuel supply problems. These are the more detailed checks that you can carry out in order to isolate faults.

FUEL STARVATION: The source of very many diesel problems is fuel starvation. To check that all is flowing well through the

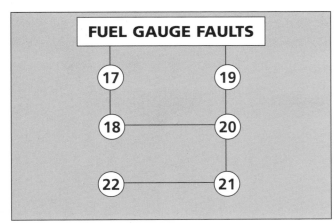

FUEL GAUGE FAULTS

17 — 19

18 — 20

22 — 21

fuel supply system, you can carry out some particularly telling checks with the aid of a vacuum gauge and suitable connection adaptors - usually banjo bolts, although this varies with the type of car.

☐ Job 1. Vacuum-testing the fuel supply circuit.

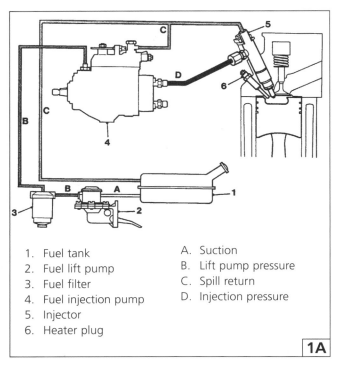

1. Fuel tank	A. Suction
2. Fuel lift pump	B. Lift pump pressure
3. Fuel filter	C. Spill return
4. Fuel injection pump	D. Injection pressure
5. Injector	
6. Heater plug	

1A

1A. It might help you to trace which pipe is which to follow this schematic view of a simple pipe layout. (Illustration, courtesy V L Churchill/LDV Limited)

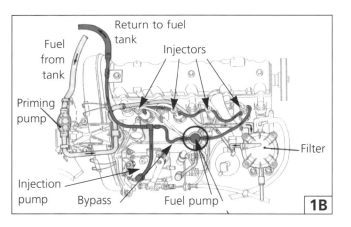

1B

1B. In practice, this gives a better idea of how the under-bonnet layout might appear. (Illustration, courtesy Peugeot)

1C

1C. The first check is of the fuel supply circuit up to the inlet of the fuel filter - in other words the tank, tank breather and outlet

gauze, and the supply pipe to the engine bay. This is done by disconnecting the fuel inlet pipe from the filter, then reconnecting it together with the vacuum gauge connection.

Start the engine and rev it up to 2,500 rpm while looking for a depression (vacuum) reading of under 1 bar (1.5 psi). If it's above that, you should check the pipes for obstruction (kinking and trapping are the usual reasons), the outlet gauze in the tank for cleanliness, and the tank breather... for its ability to breathe!). Don't forget that in the real world of physics it's not the pump that "sucks" the fuel out of the tank, it's atmospheric pressure that pushes it out - so if the breather is congested you won't get enough pressure, and that means no fuel.

1D. Next check for depression at the fuel filter outlet, providing you've established that depression at the inlet is okay. To do this disconnect the fuel outlet pipe, reconnect it together with the gauge, and again start the engine. Rev it to 2,500 rpm, verifying that depression doesn't exceed 0,25 bar (3.5 p.s.i); if it does, all you need do is renew the filter element, because it's mucky!

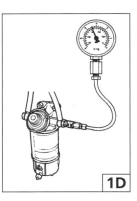

1D

1E. This is the type of fuel supply tester you will need in order to carry out the previous tests. Ensure that it has the necessary adaptors! (Illustrations, courtesy V L Churchill/LDV Limited)

Looking for air in the fuel circuit.

Now you need to look for air in the fuel circuit. This is the easy method: with the vacuum gauge still connected, wait for ten minutes to see if the depression reduces - if it does, you know there's an air leak... somewhere.

1E

1F. Another method is to disconnect the supply pipe between fuel filter (the outlet pipe) and injection pump, and fit a transparent pipe in its place. Prime and bleed the fuel system, then start the engine and run it at about 2,500 rpm. Looking at the transparent pipe you'll soon see if bubbles are coursing through it, accepting that tiny bubbles will always be present and are harmless. It's the big chaps you're on the look-out for.

If there are big bubbles, disconnect the fuel filter inlet

1F

pipe and reconnect it with a transparent pipe between the end of the inlet pipe and the filter so that, once again, you can see the fuel and/or air bubbles on the move. With the engine again running at a very fast idle, look for big bubbles in this pipe. If you see them, you know to search for the source of air entry between the filter and the tank, but if there aren't any visible you can suspect the filter assembly of being leaky. From experience, it's more often the filter assembly that is at fault than the preceding part of the system - it has more potential sources of leakage in one small area than any other part of the system.

Temporarily stopping air leaks.

1G. And that's really all you need do to check that the supply system is fundamentally okay. One final tip: if you're having trouble deciding exactly where the air is leaking in, temporarily smear thick grease around each union in turn until the bubbles stop. That should pinpoint the problem - but don't forget that the pipes themselves can be porous.

PRE-HEAT SYSTEM AND GLOW PLUG CHECKS

Because of the sheer number of differing pre-heating system types in use, it isn't possible to give detailed advice on system fault-finding, nor on pre-heating control unit diagnosis. In most cases, though, basic fault-finding can be successfully carried out as a series of systematic continuity checks starting at the ignition switch, progressing through the control unit and relays, to the glow plugs themselves.

Words of warning though: don't connect a test lamp in parallel between the glow plug warning lamp terminal at the control unit, and earth, because major damage can result to the unit through current overload. By the same token, avoid accidentally shorting the glow plug warning lamp connecting wire to earth when disconnecting or re-attaching it at the bulb. Don't be tempted to test a glow plug by connecting it straight across the battery terminals to see if it heats up, as many are rated lower than 12 volts, and they may burn out at battery voltage.

It is common to find one or more relays in the pre-heating system circuit; these are used for switching in high currents remotely with a low current switch such as the ignition switch. Relays usually have DIN-standard terminal designations, making the job of fault-finding easier: terminal 30 is the input from the positive battery terminal, 85 is the negative side of the relay actuation winding, 86 is the start of the winding (positive), and 87 is the output to the component being switched. Note though, that Japanese cars tend not to use

relays equipped with these convenient standard designations, so with those it's more a question of working things out by poking around with a test-meter.

☐ Job 2. Checking for voltage to glow plugs.

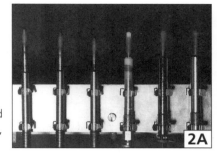

2A. This is a special test rig with a series of different types of glow plugs, some glowing more successfully than others. It's not something to be tried at home! (Illustration, courtesy Ricardo)

2B. If cold starting proves to be a problem, you should first check for voltage at the plugs with pre-heat actuated (confirmed by the pre-heat tell-tale lamp coming on at the instrument panel). Check, with the positive lead of a voltage meter held against the positive pole of the first glow plug (the one with the supply lead to it - leave it connected at this stage), and the negative lead of the meter held to earth, that the supply voltage is anything between 9 and 14 volts.

You can't expect a decent voltage reading if your battery is in a poor state of charge, so first measure battery voltage across its terminals to verify that it is at least 12 volts - any less and you know that all is not well with the battery. If all is okay on that front, but the voltage at the plugs is noticeably less than 9 volts, you can suspect an unduly high resistance in the feed circuit or the pre-heat control unit, so check all connections for looseness, dirt or corrosion.

2C. Next, check for an open-circuit (broken) glow plug element (i.e. a continuity test) with the plugs still in the cylinder head. This can be achieved by disconnecting the supply wire or strip that connects the plugs and connecting an ohmmeter (resistance meter) between the plug terminal (positive) and the screw-part of its body (earth, or negative).

2D. Or, if an ohmmeter is not to hand, by connecting a test lamp in line with the plug positive feed. If the lamp does not light up when the pre-heating system is operating, or if the ohmmeter reading is infinity, the element is broken and the plug must be renewed.

2E. Another method for ascertaining whether each glow plug is operating is to remove the injectors then look into each combustion chamber while the pre-heating system is working; the plugs should be seen to glow red after a number of seconds, but if one plug or more does not glow or is more pallid than the others, check its electrical connections then test it for continuity or high resistance with the ohmmeter. A dimly glowing test bulb indicates a high resistance. See 2A for the degree of 'glow' to expect.

INSIDE INFORMATION: Burnt glow plug elements or sheaths are usually an indication of injector faults, as damage of this type can't normally be attributed to a malfunction of the plug itself, but instead to incorrect combustion. For this reason, when there is evidence of such damage to a glow plug, its corresponding injector should always be checked for nozzle contamination and correctness of fuel spray pattern.

☐ Job 3. Checking the glow plugs current draw

A simple current-draw check on the system can tell you if one or more glow plugs are out of commission. Disconnect the glow plugs positive feed connect one connection of the ammeter to the plug; the other to the disconnected lead, operate the system and see how many amps are being drawn.

On many systems fitted to four-cylinder engines this is 48 amps - that's 12 amps consumed per plug. You will need to find the exact current rating for your car's system before carrying out this test. If, on a 48 amp system, the current is down by 12 amps, or a multiple of 12 amps, this tells you how many plugs are dead. If, for instance, the system is drawing only 24 amps, this tells you that two of the 12 amp glow plugs are inoperative.

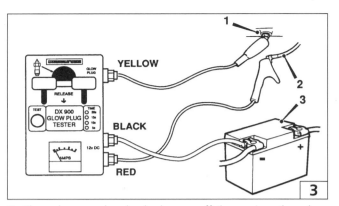

3. Glow plugs can be checked on or off the engine, though a proper inspection necessitates their removal. Purpose-built glow plug testers are available, items such as the BERU Schnelltester and the Dieseltune DX900 - see *Chapter 9, Tools & Equipment.* While these are eminently practical devices for trade use, a more commonplace multi-test meter will generally suffice for the DIY mechanic.

LOCATING A TROUBLESOME INJECTOR

A faulty injector can usually be located by listening for changes in engine tone when each injector is disabled by slackening its high pressure union. You can draw an analogy here with the system of disconnecting spark plug leads one by one on a petrol engine to find which is the 'dead' cylinder.

If there is no change in engine speed, or if one or more of the problem symptoms disappears when a union is loosened, it can be assumed that the injector in question is at fault.

CYLINDER COMPRESSION TESTING (Category: Optional/Specialist)

Compression testing is invaluable for diagnosing engine faults. A conventional (i.e. petrol-engine type) compression gauge can't be used on diesels due to the enormously high compression pressures, not to mention the absence of a spark plug hole in which to thread the gauge hose.

☐ Job 4. Using a compression tester

4A. A specific diesel compression tester has to be used, screwed or clamped into either the injector or glow plug hole of the relevant cylinder. (Illustration, courtesy V L Churchill/LDV Limited)

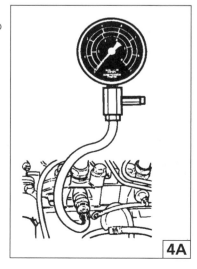

4B. SAFETY FIRST!
When testing, all injectors (or glow plugs) should be removed in order to avoid battery overload, while fuel delivery from the injection pump should be stopped by disconnecting the wire to the pump shut-off solenoid. Just undo the terminal nut on the solenoid and slip the wire off (arrowed) - but insulate it so that it doesn't short against the engine.

4C. And here's a slightly different type (connection arrowed). Ensure that you know which the correct connection is before continuing. Consult your main dealer, if necessary.

4D. Some compression gauges are designed to operate with the other injectors or plugs in place, and the engine actually idling. Where this is the case it is important to connect a length of hose between the open injector pipe and the fuel return line to avoid wastage of the fuel pumped to the inoperative cylinder. Gauges designed for dynamic use tend to

be more accurate - but they're not essential.

INSIDE INFORMATION: Unlike on a petrol engine, the accelerator doesn't need to be held open during the test as there is no throttle butterfly to restrict air flow. Having said this, there are a very few light diesels around that make use of an in-line injection pump with a pneumatic governor (some early Daihatsu off-roaders, for example) and where this is the case there is a throttle butterfly that must be held open by means of the throttle cable.

Finally, check that the air filter element is clean and that the ducting is unrestricted and that the engine is at normal operating temperature before carrying out a cylinder compression test.

CHAPTER 7
GETTING THROUGH THE MOT

Taking your beloved diesel for the annual MoT test can be rather like going to the dentist - you're not sure what to expect and the result could be painful - not only to your pocket! However, it needn't be like that...

This Chapter is for owners in Britain whose cars need to pass the 'MoT' test. The Test was first established in 1961 by the then-named Ministry of Transport: the name of the Test remains, even though the name of the government department does not!

The information in this Chapter could be very useful to non-UK owners in helping them to carry out a detailed check of a car's condition - such as when checking over a car that you might be interested in buying, for instance. But it is MOST IMPORTANT that UK owners check for themselves that legislation has not changed since this book was written and that non-UK owners obtain information on the legal requirements in their own territory - and that they act upon them. Part I gives you an overview of what the Test demands while Part II provides you with a step-by-step checklist. Check and tick every one of the check points shown here, and you'll go a long way towards passing the MoT!

Pass the MoT!

The aim of this chapter is to explain what is actually tested on a diesel car and (if it is not obvious) how the test is done. This should enable you to identify and eliminate problems before they undermine the safety or diminish the performance of your car and long before they cause the expense and inconvenience of a test failure.

The purpose of the MoT test is to try to ensure that vehicles using British roads reach minimum standards of safety. Accordingly, it is an offence to use a car without a current MoT certificate. Approximately 40 per cent of vehicles submitted for

the test fail it, but many of these failures could be avoided by knowing what the car might 'fall down' on, and by taking appropriate remedial action before the test 'proper' is carriedout. It is also worth noting that a car can be submitted for a test up to a month before the current certificate expires - if the vehicle passes, the new certificate will be valid until one year from the date of expiry of the old one, provided that the old certificate is produced at the time of the test.

The scope of the test has been considerably enlarged in the last few years, with the result that it is correspondingly more

---

SAFETY FIRST!

The MoT tester will follow a set procedure and we will cover the ground in a similar way, starting inside your car, then continuing outside, under the bonnet, underneath the car etc. When preparing to go underneath the car, do ensure that it is jacked on firm level ground and then supported on axle stands or ramps which are adequate for the task. Wheels which remain on the ground should have chocks in front of and behind them, and while the rear wheels remain on the ground, the hand

brake should be firmly ON. For most repair and replacement jobs under a car these normal precautions will suffice. However, the car needs to be even more stable than usual when carrying out these checks. There must be no risk of it toppling off its stands while suspension and steering components are being pushed and pulled in order to test them. Read carefully Chapter 1, Safety First! for further important information on raising and supporting the car above the ground.

difficult to be sure that your car will reach the required standards. In truth, however, a careful examination of the car in the relevant areas, perhaps a month or so before the current certificate expires, will highlight components which require attention, and enable any obvious faults to be rectified before you take the car for its test.

If the car is muddy or particularly dirty (especially underneath) it would be worth giving it a thorough clean a day or two before carrying out the inspection so that it has ample time to dry. Do the same before the real MoT test. A clean car makes a better impression on the examiner, who can refuse to test a car which is particularly dirty underneath.

MoT testers do not dismantle assemblies during the test but you may wish to do so during your pre-test check-up for a better view of certain wearing parts, such as the rear brake shoes for example.

Tool Box

Dismantling apart, few tools are needed for testing. A light hammer is useful for tapping panels underneath the car when looking for rust. If this produces a bright metallic noise, then the area being tapped is solid metal. If the noise produced is dull, the area contains rust or filler. When tapping sills and box sections, listen also for the sound of debris (that is, rust flakes) on the inside of the panel. Use a screwdriver to prod weak parts of panels. This may produce holes of course, but if the panels have rusted to that extent, you really ought to know about it. A strong lever (such as a tyre lever) can be useful for applying the required force to suspension joints etc. when assessing whether there is any wear in the joints. You will need an assistant to operate controls and perhaps to wobble the road wheels while you inspect components under the car.

Two more brief explanations are required before we start our informal test. Firstly, the age of the car determines exactly which lights, seat belts and other items it should have. Frequently in the next few pages you will come across the phrase "Cars first used ..." followed by a date. A car's "first used date" is either its date of first registration, or the date six months after it was manufactured, whichever was earlier. Or, if the car was originally used without being registered (such as a car which has been imported to the U.K. or an ex-H.M. Forces car etc.) the "first used date" is the date of manufacture.

Secondly, there must not be excessive rust, serious distortion or any fractures affecting certain prescribed areas of the bodywork. These prescribed areas are load-bearing parts of the bodywork within 30 cm. (12 in.) of anchorages or mounting points associated with testable items such as seat belts, brake pedal assemblies, master cylinders, servos, suspension and steering components and also body and subframe mountings. Keep this rule in mind while inspecting the car, but remember also that even if such damage occurs outside a prescribed area, it can cause failure of the test. Failure will occur if the damage is judged to reduce the continuity or strength of a main load-bearing part of the bodywork sufficiently to have an adverse effect on the braking or steering.

The following notes are necessarily abbreviated, and are for assistance only. They are not a definitive guide to all the MoT regulations. It is also worth mentioning that the varying degrees of discretion of individual MoT testers can mean that there are variations between the standards as applied. However, the following points should help to make you aware of the aspects which will be examined. Now, if you have your clipboard, checklist and pencil handy, let's make a start...

PART I: THE 'EASY' BITS

Checking these items is straightforward and should not take more than a few minutes - it could avoid an embarrassingly simple failure...

Lights

Within the scope of the test are headlights, side and tail lights, brake lights, direction indicators, and number plate lights, plus rear fog lights on all cars first used on or after 1 April, 1980, and any earlier cars subsequently so equipped, and also hazard warning lights on any car so fitted. All must operate, must be clean and not significantly damaged; flickering is also not permitted.

The switches should also all work properly. Pairs of lights should give approximately the same intensity of light output, and operation of one set of lights should not affect the working of another - such trouble is usually due to bad earthing.

Indicators should flash at between 60 and 120 times per minute. If they don't, renew the flasher unit and check all wiring and earth connections.

Interior 'tell-tale' lights, such as for indicators, rear fog lights and hazard warning lights should all operate in unison with their respective exterior lights.

Head light aim must be correct - in particular, the lights should not dazzle other road users. An approximate guide can be obtained by shining the lights against a vertical wall, but final adjustment may be necessary by reference to the beam checking machine at the MoT station. Most testers will be happy to make slight adjustments where necessary but only if the adjusters work - make sure before you take the car in that they are not seized solid! Reflectors must be unbroken, clean, and not obscured - for example, by stickers.

Wheels and Tyres

Check the wheels for loose nuts/bolts, cracks, and damaged rims. Missing wheel nuts or studs are also failure points, naturally enough!

There is no excuse for running on illegal tyres. The legal requirement is that there must be at least 1.6 mm. of tread depth remaining, over the 'central' three-quarters of the width of the tyre all the way around. From this it can be deduced that there is no legal requirement to have 1.6 mm. (1/16 in.) of tread on the 'shoulders' of the tyre, but in practice, most MoT stations will be reluctant to pass a tyre in this condition. In any case, for optimum safety - especially 'wet grip' - you would be well advised to change tyres when they wear down to around 3 mm. (1/8 in.) or so depth of remaining tread.

Visible 'tread wear indicator bars', found approximately every nine inches around the tread of the tyre, are highlighted when the tread reaches the critical 1.6 mm. point.

Tyres should not show signs of cuts or bulges, evidence of rubbing on the bodywork or running gear, and the valves should be in sound condition, and correctly aligned.

Cross-ply and radial tyre types must not be mixed on the same axle (i.e. the front or rear of the car), and if pairs of cross-ply and radial tyres are fitted, the radials must be on the rear axle (rear end).

Windscreen

The screen must not be damaged (by cracks, chips, etc.) or obscured so that the driver does not have a clear view of the road. See Inside the *Car, Check 4.* below.

Washers and Wipers

The wipers must clear an area big enough to give the driver a clear view forwards and to the side of the car. (In other words, one wiper won't do!) The wiper blades must be securely attached and sound, with no cracks or 'missing' sections. The wiper switch should also work properly. The screen washers must supply the screen with sufficient liquid to keep it clean, in conjunction with the use of the wipers.

Mirrors

If your car was first used before 1 August 1978, it doesn't need by law to have an exterior mirror in addition to its interior mirror. Later cars must have at least two mirrors, one of which must be on the driver's side. The mirrors must be visible from the driver's seat, and not be damaged or obscured so that the view to the rear is affected. Therefore cracks, chips and discolouration can mean failure.

Horn

The horn must emit a uniform note which is loud enough to give adequate warning of approach, and the switch must operate correctly. Multi-tone horns playing 'in sequence' are not permitted, but two tones sounding together are fine.

Seat Security

The seats must be securely mounted, and the sub-frames should be sound. Seat recliners must lock securely in position.

Number (Registration) Plates

Both front and rear number plates must be present, and in good condition, with no breaks or missing numbers or letters. The plates must not be obscured, and the digits must not be repositioned (to form names, for instance).

Vehicle Identification Numbers (VIN)

Cars first used on or after 1 August 1980 are obliged to have a clearly displayed VIN - Vehicle Identification Number (or old-fashioned 'chassis number' for older cars), which is plainly legible.

Exhaust System

The entire system must be present, properly mounted, free of leaks and should not be noisy - which can happen when the internal baffles fail. 'Proper' repairs by welding, or exhaust cement, or bandage are acceptable, as long as no gas leaks are evident. Then again, common sense, if not the MoT, dictates that exhaust bandage should only be a very short-term emergency measure. For safety's sake, fit a new exhaust if yours is reduced to this!

Seat Belts

Belts are not needed on cars first used before 1 January, 1965. On cars after this date - and earlier examples, if subsequently fitted with seat belts - the belts must be in good condition (i.e. not frayed or otherwise damaged), and the buckles and catches should also operate correctly. Inertia reel types, where fitted, should retract properly.

Belt mountings must be secure, with no structural damage or corrosion within 30 cm. (12 in.) of them.

Part II: MORE DETAILED STUFF

You've checked the easy bits - now it's time for the detail! Some of the 'easy bits' referred to above are included here, but this is intended as a more complete check list to give your car the best possible chance of gaining a First Class Honours, MoT Pass!

Inside The Car

☐ 1. The steering wheel should be examined for cracks and for damage which might interfere with its use, or injure the driver's hands. It should also be pushed and pulled along the column axis, and also up and down, at 90 degrees to it. This will highlight any deficiencies in the wheel and the upper column mounting/bearing, and also any excessive end float, and movement between the column shaft and the wheel. Rotate the steering wheel in both directions to test for free play at the wheel rim - this shouldn't exceed approximately 13mm (1/2 in.) a the rim, assuming a 380 mm. (15 in.) diameter steering wheel on a car with steering rack-type steering. Free movement on (the majority of) cars with a smaller wheel is proportionally less: this is the figure quoted by the Department of Transport. Look, too, for movement in the steering column couplings and fasteners (including the universal joint, when the column has one), and visually check their condition and security. They must be sound, and properly tightened.

☐ 2. Check that the switches for headlights, sidelights, direction indicators, hazard warning lights, wipers, washers and horn, appear to be in good working order and check that the tell-tale lights or audible warnings are working where applicable.

☐ 3. Make sure that the windscreen wipers operate effectively with blades that are secure and in good condition. The windscreen washer should provide sufficient liquid to clear the screen in conjunction with the wipers.

☐ 4. Check for windscreen damage, especially in the area swept by the wipers. From the MoT tester's point of view, Zone A is part of this area, 290 mm. (11.5 in.) wide and centred on the centre of the steering wheel. Damage to the screen within this area should be capable of fitting into a

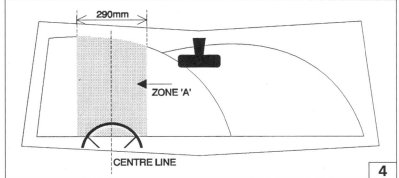

10 mm. (approx. 0.4 in.) diameter circle and the cumulative effect of more minor damage should not seriously restrict the driver's view. Windscreen stickers or other obstructions should not encroach more than 10 mm. (approx 0.4 in.) into this area. In the remainder of the area swept by the wiper, the maximum diameter of damage or degree of encroachment by obstructions is 40 mm. (approx. 1.6 in.) and there is no ruling regarding cumulative damage. Specialist windscreen companies can often repair a cracked screen for a lot less than the cost of replacement. Moreover, the cost of repair is often covered by comprehensive insurance policies without excess but check first that your no-claims discount won't suffer!

☐ 5. The horn control should be present, secure and readily accessible to the driver, and the horn should be loud enough to be heard by other road users. Gongs, bells and sirens are not permitted (except as part of an anti-theft device) and multi- tone horns (which alternate between two or more notes) are not permitted at all. On cars first used after 1 August 1973, the horn should produce a constant, continuous or uniform note which is neither harsh nor grating.

☐ 6. After 1 August 1978 there must be one exterior mirror on the driver's side of the vehicle and either an exterior mirror fitted to the passenger's side or an interior mirror. The required mirrors should be secure and in good condition.

☐ 7. Check that the handbrake operates effectively without coming to the end of its working travel. The lever and its mechanism must be complete, securely mounted, unobstructed in its travel and in a sufficiently good condition to remain firmly in the "On" position even when knocked from side to side. The rule regarding bodywork corrosion applies in the vicinity of the handbrake lever mounting: there must be none within 30 cm (12 in.).

☐ 8. The foot brake pedal assembly should be complete, unobstructed, and in a good working condition, including the pedal rubber (which should not have been worn smooth). There should be no excessive movement of the pedal at right angles to its normal direction. When fully depressed, the pedal should not be at the end of its travel. The pedal should not feel spongy (indicating air in the hydraulic system), nor should it tend to creep downwards while held under pressure (which indicates an internal hydraulic leak).

☐ 9. Seats must be secure on their mountings and seat backs must be capable of being locked in the upright position. However, if your car is of an age where it was not part of the original design for the seat back to be locked, it will not be an MoT failure point. (This applies to virtually no diesel engined cars.)

☐ 10. On cars first used on or after 1 January 1965, but before 1 April 1981, the driver's and front passenger's seats must be fitted with seat belts, though these can be simple diagonal belts rather than the three-point belts (lap and diagonal belts for adults with at least three anchorage points) required by later cars. For safety's sake, however, we do not recommend this type of belt. Rear seat belts are a requirement for cars first used after 31 March 1987. Examine seat belt webbing and fittings to make sure that all are in good condition and that anchorages are firmly attached to the car's structure. Locking mechanisms should be capable of remaining locked, and of being released if required, when under load. Flexible buckle stalks (if fitted) should be free of corrosion, broken cable strands or other weaknesses. Note that any belts fitted which are not part of a legal requirement may be examined by the tester but will not form part of the official test.

☐ 11. On inertia reel belts, check that on retracting the belts the webbing winds into the retracting unit automatically, albeit with some manual assistance to start with.

☐ 12. Note the point raised earlier regarding corrosion around seat belt anchorage points. The MoT tester will not carry out any dismantling here, but he will examine floor mounted anchorage points from underneath the car if that is possible.

☐ 13. Before getting out of the car, make sure that both doors can be opened from the inside.

Outside The Car

☐ 14. Before closing the driver's door check the condition of the inner sill. Usually the MoT tester will do this by applying finger or thumb pressure to various parts of the panel while the floor covering remains in place. For your own peace of mind, look beneath the sill covering, taking great care not to tear any rubber covers. Then close the driver's door and make sure that it latches securely and repeat these checks on the nearside inner sill and door.

Now check all of the lights, front and rear, (and the number plate lights) while your assistant operates the light switches. Remember - before you panic - that brake lights (and indicators) only work with the ignition turned on!

☐ 15. As we said earlier, you can carry out a rough and ready check on head light alignment for yourself, although it will certainly not be as accurate as having it done for you at the MoT testing station. Drive your car near to a wall, as shown. Check that your tyres are correctly inflated and the car is on level ground.

Draw on the wall, with chalk:
• a horizontal line about 2 metres long, and at same height as centre of head lamp lens.
• two vertical lines about 1 metre long, each forming a cross with the horizontal line and the same distance apart as the head lamp centres.
• another vertical line to form a cross on the horizontal line, midway between the others.

Now position your car so that:
• it faces the wall squarely, and its centre line is in line with centre line marked on the wall.
• the steering is straight.
• head light lenses are 5.0 metres (16 ft) from the wall.

Switch on the headlights' 'main' and 'dipped' beams in turn, and measure their centre points. You'll be able to see them better at night, of course. You will be able to judge any major discrepancies in intensity and aim prior to having the beams properly set by a garage with beam measuring equipment.

Headlights should be complete, clean, securely mounted, in good working order and not adversely affected by the operation of another light, and these basic requirements affect all the lights listed below. Headlights must dip as a pair from a single switch. Their aim must be correctly adjusted and they should not be affected (even to the extent of flickering) when lightly tapped by hand. Each head light should match its partner in terms of size, colour and intensity of light, and can be white or yellow but we repeat - both must be the same.

☐ 16. Side lights should show white light to the front and red light to the rear. Lenses should not be broken, cracked or incomplete.

☐ 17. Vehicles first used before 1 April 1986 do not have to have a hazard warning device, but if one is fitted, it must be tested, and it must operate with the ignition switch in both on and off positions. The lights should flash 60-120 times per minute, and indicators must operate independently of any other lights.

☐ 18. Check your stop lights. Pre-1971 cars need only one, but when two are fitted, both are tested, so you will not get away with one that works and one that doesn't! Stop lights should produce a steady red light when the foot brake is applied.

☐ 19. There must be two red rear reflectors - always fitted by the manufacturers, of course! - which are clean, and securely and symmetrically fitted to the car.

☐ 20. Cars first used on or after 1 April 1980 must have one rear fog light fitted to the centre or offside of the car and, as far as fog lights are concerned, the MoT tester is interested in this light on these cars only. It must comply with the basic requirements (listed under headlights) and emit a steady red light. Its tell-tale light, inside the car, must work to inform the driver that it is switched on.

☐ 21. There must be a registration number plate at the front and rear of the car and both must be clean, secure, complete and unobscured. Letters and figures must be correctly formed and correctly spaced and not likely to be misread due to an

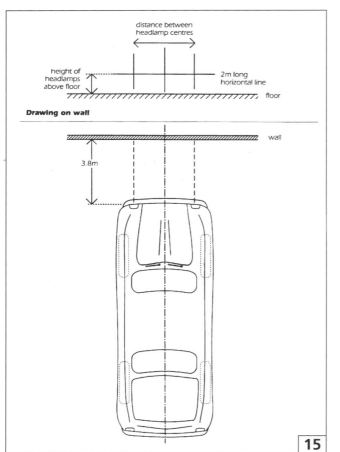

distance between headlamp centres

height of headlamps above floor

2m long horizontal line

floor

Drawing on wall

wall

3.8m

15

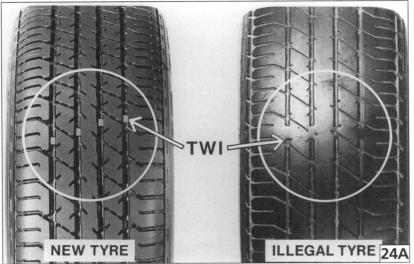

NEW TYRE   ILLEGAL TYRE   **24A**

The following six photographs and information in this section have been supplied with grateful thanks to Dunlop/SP Tyres.

☐ 24A. Modern tyres have tread wear indicators built into the tread grooves (usually about eight of them spread equidistantly around the circumference). These appear as continuous bars running across the tread when the original pattern depth has worn down to 1.6 mm. There will be a distinct reduction in wet grip well before the tread wear indicators start to show, and we recommend that you should replace tyres before they get to this stage, even though this is the legal minimum in the UK.

24B

uncovered securing bolt or whatever. The year letter counts as a figure. The space between letters and figures must be at least twice that between adjacent letters or figures.

☐ 22. Number plate lights must be present, working, and must not flicker when tapped by hand, just as for other lights. Where more than one light or bulb was fitted as original equipment, all must be working. The MoT tester will examine tyres and wheels while walking around the car and again when he is under the car.

☐ 23. Front tyres should match each other and rear tyres should match each other, both sets matching in terms of size, aspect ratio and type of structure. For example, you must never fit tyres of different sizes or types, such as cross-ply or radial, on the same 'axle' - both front wheels counting as 'on the same axle' in this context. Cross-ply or bias belted tyres should not be fitted on the rear axle, with radials on the front, neither should cross-ply tyres be fitted to the rear, with bias belted tyres on the front.

☐ 24. Failure of the test can be caused by a cut, lump, tear or bulge in a tyre, exposed ply or cord, a badly seated tyre, a re-cut tyre, a tyre fouling part of the vehicle, or a seriously damaged or misaligned valve stem which could cause sudden deflation of the tyre. To pass the test, the grooves of the tread pattern must be at least 1.6 mm. deep throughout a continuous band comprising the central three-quarters of the breadth of tread, and round the entire outer circumference of the tyre.

☐ 24B. Lumps and bulges in the tyre wall usually arise from accidental damage or even because of faults in the tyre construction. You should run your hand all the way around the side wall of the tyre, with the car either jacked off the ground, or moving the car half a wheel's revolution, so that you can check the part of the tyre that was previously resting on the ground. Since you can't easily check the insides of the tyres in day-to-day use, it is even more important that you spend time carefully checking the inside of each tyre - the MoT tester will certainly do so! Tyres with bulges in them must be scrapped and replaced with new, since they can fail suddenly, causing your car to lose control.

☐ 24C. Abrasion of the tyre side wall can take place either in conjunction with bulging, or by itself, and this invariably results from an impact, such as the tyre striking the edge of a kerb or a pothole in the road. Once again, the tyre may be at imminent risk of failure and you should take advice from a tyre specialist on whether the abrasion is just superficial, or whether the tyre will need replacement.

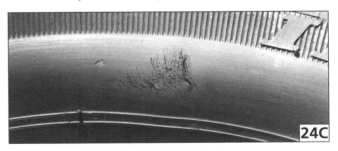

24C

☐ 24D. All tyres will suffer progressively from cracking, albeit in most cases superficially, due to the effects of ozone and

24D

sunlight. If old age has caused the tyres on your car to degrade to this extent, replace them.

□ 24E. If the outer edges of the tread are worn noticeably more than at the centre, the tyres have been run under-inflated, which not only ruins tyres, but causes worse fuel consumption, dangerous handling and is, of course, illegal.

24E

Over-inflation causes the centre part of the tyre to wear more quickly than the outer edges. This is also illegal but in addition, it causes the steering and grip to suffer and the tyre becomes more susceptible to concussion damage.

□ 24F. Incorrect wheel alignment causes one side of the tyre to wear more severely than the other. If your car should ever hit a kerb or large pothole, it is worthwhile having the wheel alignment checked since this costs considerably less than new front tyres!

24F

□ 25. Road wheels must be secure and must not be badly damaged, distorted or cracked, or have badly distorted bead rims (perhaps due to "kerbing"), or loose or missing wheel nuts, studs or bolts.

□ 26. Check the bodywork for any sharp edges or projections, caused by corrosion or damage, which could prove dangerous to other road users, including pedestrians.

□ 27. Check that the fuel cap fastens securely and that its sealing washer is neither torn nor deteriorated, or its mounting flange damaged sufficiently to allow fuel to escape (for example, while the car is cornering).

Under The Bonnet

□ 28. The car should have a Chassis Number or Vehicle Identification Number fitted to the bodywork. The modern VIN plate is required on all vehicles first used on or after 1 August 1980. This can be on a plate secured to the vehicle or, etched or stamped on the bodywork.

□ 29. While peering under the bonnet check that the brake hydraulic reservoir is securely mounted and not damaged. Ensure that the cap is present, that the fluid levels are satisfactory and that there are no fluid leaks.

□ 30. Still under the bonnet, make a thorough search for evidence of excessive corrosion, severe distortion or fracture in any load bearing panelling within 30cm (12 in.) of important mounting points such as the master cylinder and/or servo mounting, front suspension mountings etc.

Under The Car - Front End

SAFETY FIRST!
On some occasions there is no alternative but for your assistant to sit in the car whilst you go beneath. Therefore:
1) Place the car ramps as well as axle stands beneath the car's structure so that it cannot fall.
2) Don't allow your assistant to move vigorously or get in or out of the car while you are beneath it. If either of these are problematical, DON'T CARRY OUT THIS CHECK - leave it to your garage.

□ 31. Have an assistant turn the steering wheel from side to side while you watch for movement in the steering mechanism. Make sure that the steering rack is securely clamped to car body or subframe, that the joints show no signs of wear and that the ball joint dust covers are in sound condition. Ensure that all split pins, locking nuts and so on are in place and correctly fastened, throughout the steering and suspension systems.

□ 32. Check the steering column universal joint and clamp bolts (where applicable) for wear and looseness by asking your assistant to turn the steering wheel from side to side while you watch what happens to the wheels. More than 13mm (approx. 1/2 in.) of free play at the perimeter of the steering wheel (cars with a steering rack) or more than 75mm (3 in.) (cars with a box-type steering), due to wear in the steering components, is sufficient grounds for a test failure. Note that the 75mm criterion is based on a steering wheel of 380mm (15 in.) diameter and will be less for smaller steering wheels. Also check for the presence and security of retaining and locking devices in the steering column assembly (depending on year and model).

□ 33. Examine the suspension springs (and their attachments) to ensure that they are free from cracks, fractures and excessive corrosion and pitting. In addition, check the end fixings for security and excessive free play.

□ 34. With the front of the car raised, and securely supported on axle stands (with the wheels clear of the ground), grasp each front wheel/tyre in turn at top and bottom, and attempt to 'rock' the wheel in and out. If more than just perceptible movement is evident at the rim, this could be due to wear in the hub swivel ball joint(s), or in the front wheel bearings. Repeat the test while an assistant applies the foot brake. This will effectively lock the hub to the stub axle assembly, so any movement remaining will be in the hub swivel(s).

☐ 35. Check the condition of the master cylinder, positioned immediately behind the brake pedal, on the other side of the bulkhead. It should be securely mounted and show no signs of leaks. Make sure that any electrical connections to it are sound.

☐ 36. Spin each front wheel in turn, listening for roughness in the bearings. Don't be misled by the drag of any axle shafts. There must be no roughness.

☐ 37. Visually inspect the shock absorbers for signs of leaks or corrosion. Don't confuse road spray with leaking hydraulic oil. Where fitted, coil springs must be sound and free from cracks or other visible damage.

☐ 38. With the wheels on the ground again, push down firmly a couple of times on each front wing of the car, then let go at the bottom of a stroke. The car should return to approximately its original level within two or three strokes. Do the same to the rear of the car. Continuing oscillations will earn your car a 'failure' ticket for worn 'shockers'!

Under The Car - Rear Suspension

☐ 39. Check the condition of the rear shock absorbers in the same way as the front. A 'bounce' test can be carried out as for the front shock absorbers as an approximate check on how efficient or otherwise the damping effect is!

☐ 40. With the back of the car raised on axle stands (both rear wheels off the ground), rotate the rear wheels and check, as well as you can, for roughness in the bearings, just as you did at the front. Once again, don't be misled by the 'drag' from any axle shafts.

Brakes

☐ 41. The MoT brake test is carried out on a special 'rolling road' set-up, which measures the efficiency in terms of percentage. For the foot brake, the examiner is looking for 50 per cent; the handbrake must measure 25 per cent. Frankly, without a rolling road of your own, there is little that you can do to verify whether or not your car will come up to the required figures. What you can do, though, is carry out an entire check of the brake system, which will also cover all other aspects the examiner will be checking, and be as sure as you can that the system is working efficiently.

☐ 42. The MoT examiner will not dismantle any part of the system, but you can do so. So, take off each front wheel in turn, and examine as follows:

Disc Brakes

SAFETY FIRST! AND SPECIALIST SERVICE.
Obviously, your car's brakes are among its most important safety related items. Do NOT dismantle or attempt to perform any work on the braking system unless you are fully competent to do so. If you have not been trained in this work, but wish to carry it out, we strongly recommend that you have a garage or qualified mechanic check your work before using the car on the road. See also the section on BRAKES AND ASBESTOS in Chapter 1, for further information. Always replace the disc pads and/or shoes in sets of four - never replace the pads/shoes on one wheel only.

☐ 43. Check the front (or rear) brake discs themselves, looking for excessive grooving or crazing, the caliper pistons and dust seals (looking for signs of fluid leakage and deterioration of the seals), and the brake pads - ideally, replace them if less than approximately 3 mm. (1/8 in.) friction material remains on each pad.

Drum Brakes

☐ 44. Remove each brake drum and check the condition of the linings (renew if worn down to anywhere near the rivet heads), the brake drum (watch for cracking, ovality and serious scoring, etc.) and the wheel cylinders. Check the cylinder's dust covers to see if they contain brake fluid. If so, or if it is obvious that the cylinder(s) have been leaking, in which case they require replacement or overhaul.

☐ 45. Ensure that drum brake adjusters (where fitted - older cars) are free to rotate (i.e. not seized!). If they are stuck fast, apply a little penetrating oil to the backs of the adjusters, taking great care not to get any onto the brake shoes. Work them gently until they are free, at which point a little brake grease can be applied to the threads to keep them in this condition. With the rear brakes correctly adjusted, check the handbrake action. The lever should rise by a specified number of 'clicks' (see your model-specific manual) before the brake operates fully - if it goes further, the cable or rear brakes require adjustment. Ensure too that the handbrake lever remains locked in the 'on' position when fully applied, even if the lever is knocked sideways.

☐ 46. Closely check the state of ALL visible pipework - that's brakes and fuel lines. If any section of the steel tubing shows signs of corrosion, replace it, for safety as well as to gain an MoT pass. Note also that where the manufacturer originally fitted a clip to secure a piece of pipe, it must be present and the pipe must be secured by it. Look too for leakage of fluid around pipe joints, and from the master cylinder. The fluid level in the master cylinder reservoir must also be at its correct level - if not, find out why and rectify the problem! At the front and rear of the car, bend the flexible hydraulic pipes through 180 degrees (by hand) near each end of each pipe, checking for signs of cracking. If any is evident, or if the pipes have been chafing on the tyres, wheels, steering or suspension components, replace them with new items, re-routing them to avoid future problems.

☐ 47. Have an assistant press down hard on the brake pedal while you check all flexible pipes for bulges. As an additional check, firmly apply the foot brake and hold the pedal down for a few minutes. It should not slowly sink to the floor (if it does, you have a hydraulic system problem). Press and release the pedal a few times - it should not feel 'spongy' (due to the presence of air in the system). If there is the risk of any problems with the braking system's hydraulics, have a qualified mechanic check it over before using the car.

☐ 48. A test drive should reveal obvious faults (such as pulling to one side, due to a seized caliper piston, for example), but otherwise all will be revealed on the rollers at the MoT station...

Bodywork Structure

☐ 49. A structurally deficient car is a dangerous vehicle, and rust can affect many important structural areas of the bodyshell or chassis - if the car has a separate chassis, as do vans and Land Rovers. Examine all these areas, particularly those adjacent to subframe, suspension, steering, brake component and power unit mountings. Also the spare wheel well and the roadside jack mounting points.

☐ 50. Essentially, fractures, cracks or serious corrosion in any load bearing panel or member (to the extent that the affected sections are weakened) need to be dealt with. In addition, failure will result from any deficiencies in the structural metalwork within 30 cm. (12 in.) of the seat belt mountings, and also the steering and suspension component attachment points. Repairs made to any structural areas must be carried out by 'continuous' seam welding, and the repair should restore the affected section to at least its original strength.

☐ 51. The MoT examiner will be looking for metal which gives way under squeezing pressure between finger and thumb, and will use his wicked little 'Corrosion Assessment Tool' (i.e. a plastic-headed tool, known as the 'toffee hammer'[1]), which in theory at least should be used for detecting rust by lightly tapping the surface. If scraping the surface of the metal shows weakness beneath, the car will fail.

☐ 52. Note that the security of doors and other openings must also be assessed, including the hinges, locks and catches. Corrosion damage or other weakness in the vicinity of these items can mean failure. It must be possible to open both doors from inside and outside the car.

Exterior Bodywork

☐ 53. Check for another area which can cause problems. Look out for surface rust, or accident damage on the exterior bodywork, which leaves sharp/jagged edges and which may be liable to cause injury. Ideally, repairs should be carried out

by welding in new metal, but for non-structural areas, riveting a plate over a hole, bridging the gap with glass fibre or body filler, or even taping over the gap can be legally acceptable, at least as far as the MoT test is concerned.

Fuel System

☐ 54. Another recent extension of the regulations brings the whole of the fuel system under scrutiny, from the tank to the engine. The system should be examined with and without the engine running, and there must be no leaks from any of the components. The tank must be securely mounted, and the filler cap must fit properly - 'temporary' caps are not permitted.

Emissions

In almost every case, a proper check on engine running (as detailed in *Chapter 5 - Diesel Servicing*) will help to ensure that your car is running at optimum efficiency, and there should be no difficulty in passing the test, unless your engine or the injection system are well worn. Before submitting your diesel for its MoT test, make sure the engine is at normal operating temperature, and if you have any doubts about the age, tension or condition of its camshaft drive belt (where applicable) get this renewed BEFORE the test. This is because the engine will be given quite a hard time during the test and if a stressed cambelt snaps, the engine may be ruined - at your expense!

☐ 55. In almost every case, a proper 'engine tune' will help to ensure that your vehicle is running at optimum efficiency, and there should be no difficulty in passing the test, unless your engine is well worn or is out of adjustment. Too high a smoke output can often be cured by carrying out simple servicing procedures as described in *Chapter 5, Diesel Servicing.* It's important to ensure that the fuel is being burnt when and where it should be, which means the air filter condition, injection timing and injector performance should be tip-top. However, if smoke output is very substantial, it points to there being a serious engine problem, and professional help should be sought.

All engines are subject to the 'visual smoke emission' test. The engine must be fully warmed up, allowed to idle, then revved slightly. If smoke emitted is regarded by the examiner as being 'excessive', the vehicle will fail. In practice, attitudes vary widely between MoT stations on this aspect of the test.

☐ 56. Readings on a diesel engine which are too high to pass the test will require the attention of a *SPECIALIST SERVICE.* As we said earlier, too high a smoke output can often be cured by carrying out simple servicing procedures as described in *Chapter 5.* It's important to ensure that the fuel is being burnt when and where it should be, which means the air filter condition, injection timing and injector performance should be tip-top. However, if smoke output is very substantial, it points to there being a serious engine problem, and professional help should be sought.

FACT FILE: DIESEL ENGINES' EMISSIONS TEST

The Tester will have to rev your engine hard, several times. If it is not in good condition, he is entitled to refuse to test it. This is the full range of tests, even though all may not apply to your car.

Vehicles first used before 1 August, 1979

Engine run at normal running temperature; engine speed taken to around 2500 rpm (or half governed max. speed, if lower) and held for 20 seconds. FAILURE, if engine emits dense blue or black smoke for next 5 seconds, at tick-over. (NOTE: Testers are allowed to be more lenient with pre-1960 vehicles.)

Vehicles first used on or after 1 August, 1979

After checking engine condition, and with the engine at normal running temperature, the engine will be run up to full revs between three and six times to see whether your engine passes the prescribed smoke density test. (For what it's worth - 2.5k for non-turbo cars; 3.0k for turbo diesels. An opacity meter probe will be placed in your car's exhaust pipe and this is not something you can replicate at home.) Irrespective of the meter readings, the car will fail if smoke or vapour obscures the view of other road users.

IMPORTANT NOTE: The diesel engine test puts a lot of stress on the engine. It is IMPERATIVE that your car's engine is properly serviced, and the cam belt changed on schedule, before you take it in for the MoT test. The tester is entitled to refuse to test the car if he feels that the engine is not in serviceable condition and there are a number of pre-Test checks he may carry out.

Note that the tester is expected to check the engine oil level, check the camshaft drive belt condition and tension, where this is possible, and generally be on the alert for engine defects. If he is aware of any, he is expected to fill in a Notification of Refusal to test. The reason for this is that the engine must be revved hard during smoke testing, and some diesel engines were wrecked when this aspect of the test was originally introduced in 1992. In many cases failure of the camshaft drive belt caused internal engine damage. Since then precautions have been built into the test.

CHAPTER 8 - FACTS AND FIGURES

You'll find that the information shown here will be invaluable when it comes to servicing your car at home. It is a compilation of all the settings and capacities that you will be likely to need - although we have not included the information on jobs that can only be carried out by a specialist. He will have his own list to consult. Also, it would clearly be impossible to cover every diesel engine ever built! If your car belongs to one of the minority categories we've left room for you to fill in the missing spaces, having carried out your own research, at the end of this chapter.

AUDI

Make/model Audi 80D 1.6

Year 1982-86

Engine (code)	**1588cc (JK)**
Power	**55bhp**
Compression pressure	**6-34bar**
Firing order	**1342**
No.1 cyl	**@ cambelt end**
Valve clearances	**(hot)**
• inlet	**0.20-0.25mm**
• exhaust	**0.40-0.45mm**
Inj. pump type	**Bosch**
Idle speed	**925-975rpm**
Max. no-load speed	**5050-5150rpm**
Glow plug voltage	**11**
Pre-heat duration	**7-10secs approx**
Engine oil viscosity	**15W/40 or 50**
Oil quality	**API CC or CD**
Oil capacity (inc. filter)	**3.5 litres**
Gearbox oil grade	**SAE 80W**
Oil capacity	**1.7 litres (2.0 litres - 5speed)**
Final drive oil grade	**N/A**
Oil capacity	**N/A**
Cooling system capacity	**7.0 litres**

Make/model Audi 80D 1.6

Year 1986-89

Engine (code)	**1588cc (JK)**
Power	**55bhp**
Compression pressure	**26-34bar**
Firing order	**1342**
No.1 cyl	**@ cambelt end**
Valve clearances	**(hydraulic - no adjustment)**
• inlet	**-**
• exhaust	**-**
Inj. pump type	**Bosch**
Idle speed	**850-900rpm**
Max. no-load speed	**5300-5400rpm**
Glow plug voltage	**11**
Pre-heat duration	**7-10secs approx**
Engine oil viscosity	**15W/40 or 50**
Oil quality	**API CC or CD**
Oil capacity (inc. filter)	**3.5 litres**
Gearbox oil grade	**SAE 80W**
Oil capacity	**1.7 litres (2.0 litres - 5speed)**
Final drive oil grade	**N/A**
Oil capacity	**N/A**
Cooling system capacity	**7.0 litres**

Make/model Audi 80 TD 1.6

Year 1982-93

Engine (code)	**1588cc (CY,RA,SB)**
Power	**80bhp (70bhp - CY)**
Compression pressure	**26-34bar**
Firing order	**1342**
No.1 cyl	**@ cambelt end**
Valve clearances	**(hot) (hyd. -86 on)**
• inlet	**0.25mm**
• exhaust	**0.45mm**

Inj. pump type	Bosch or Lucas
Idle speed	925rpm
Max. no-load speed	5100rpm
Glow plug voltage	11
Pre-heat duration	7-10secs approx
Engine oil viscosity	15W/40 or 50
Oil quality	API CC or CD
Oil capacity (inc. filter)	3.5 litres
Gearbox oil grade	SAE 80W
Oil capacity	2.4 litres (2.0 litres -CY eng.)
Final drive oil grade	N/A
Oil capacity	N/A
Cooling system capacity	7.0 litres

Make/model Audi 80D 1.9

Year 1989-93

Engine (code)	1896cc (1Y)
Power	69bhp
Compression pressure	26-34bar
Firing order	1342
No.1 cyl	@ cambelt end
Valve clearances	(hydraulic - no adjustment)
• inlet	-
• exhaust	-
Inj. pump type	Bosch
Idle speed	875-925rpm
Max. no-load speed	5050-5150rpm
Glow plug voltage	11
Pre-heat duration	7-10secs approx
Engine oil viscosity	15W/40 or 50
Oil quality	API CC or CD
Oil capacity (inc. filter)	4.5 litres
Gearbox oil grade	SAE 80W
Oil capacity	2.0 litres
Final drive oil grade	N/A
Oil capacity	N/A
Cooling system capacity	7.0 litres

Make/model Audi 80TD 1.9

Year 1991-94

Engine (code)	1896cc (AAZ)
Power	75bhp
Compression pressure	26-34bar
Firing order	1342
No.1 cyl	@ cambelt end
Valve clearances	(hydraulic - no adjustment)
• inlet	-
• exhaust	-

Inj. pump type	Bosch
Idle speed	875-925rpm
Max. no-load speed	5100rpm
Glow plug voltage	11
Pre-heat duration	7-10secs approx
Engine oil viscosity	15W/40 or 50
Oil quality	API CD
Oil capacity (inc. filter)	4.5 litres
Gearbox oil grade	SAE 80W
Oil capacity	2.4 litres
Final drive oil grade	N/A
Oil capacity	N/A
Cooling system capacity	7.0 litres

CITROEN

Make/model Citroen AX 14D

Year 1988-93

Engine (code)	1360cc (TUD3)
Power	53bhp
Compression pressure	-
Firing order	1342
No.1 cyl	@ gearbox end
Valve clearances	(cold)
• inlet	0.15mm
• exhaust	0.30mm
Inj. pump type	Bosch or Lucas
Idle speed	800rpm
Max. no-load speed	5500rpm
Glow plug voltage	11
Pre-heat duration	6-10secs approx
Engine oil viscosity	15W/40
Oil quality	API CC or CD
Oil capacity (inc. filter)	3.75 litres
Gearbox oil grade	SAE 75W/80
Oil capacity	2.0 litres
Final drive oil grade	N/A
Oil capacity	N/A
Cooling system capacity	4.8 litres

FACTS AND FIGURES

Make/model Citroen BX 17D/17D Turbo

Year **1984-9**

Engine (code)	1769cc (XUD7/T)
Power	60bhp/90bhp
Compression pressure	28bar approx
Firing order	1342
No.1 cyl	@ gearbox end
Valve clearances	(cold)
• inlet	0.10mm (0.15 -Turbo)
• exhaust	0.25mm (0.15 -Turbo)
Inj. pump type	Bosch or Lucas
Idle speed	800rpm
Max. no-load speed	5100rpm (4800 -Turbo)
Glow plug voltage	11
Pre-heat duration	6-10secs approx
Engine oil viscosity	15W/40 or 50
Oil quality	API CD
Oil capacity (inc. filter)	5.0 litres
Gearbox oil grade	SAE 75W/80
Oil capacity	2.0 litres
Final drive oil grade	N/A
Oil capacity	N/A
Cooling system capacity	6.5 litres

Make/model Citroen BX 19D

Year **1984-93**

Engine (code)	1905cc (XUD9)
Power	64bhp/71bhp
Compression pressure	28bar approx
Firing order	1342
No.1 cyl	@ gearbox end
Valve clearances	(cold)
• inlet	0.15mm
• exhaust	0.30mm
Inj. pump type	Bosch or Lucas
Idle speed	800rpm
Max. no-load speed	5100rpm
Glow plug voltage	11
Pre-heat duration	6-10secs approx
Engine oil viscosity	15W/40 or 50
Oil quality	API CC or CD
Oil capacity (inc. filter)	5.0 litres
Gearbox oil grade	SAE 75W/80
Oil capacity	2.0 litres
Final drive oil grade	N/A
Oil capacity	N/A
Cooling system capacity	6.5 litres

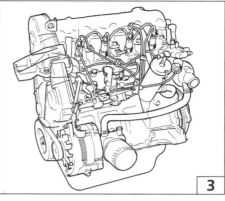

3

Make/model Citroen ZX D 1.8

Year **1991-94**

Engine (code)	1769cc (XUD7)
Power	60bhp
Compression pressure	25-30bar
Firing order	1342
No.1 cyl	@ gearbox end
Valve clearances	(cold)
• inlet	0.15mm
• exhaust	0.30mm
Inj. pump type	Bosch or Lucas
Idle speed	750-800rpm
Max. no-load speed	5000-5200rpm
Glow plug voltage	11
Pre-heat duration	7-10secs approx
Engine oil viscosity	15W/40
Oil quality	API CC or CD
Oil capacity (inc. filter)	5.0 litres
Gearbox oil grade	SAE 75W/80
Oil capacity	2.0 litres
Final drive oil grade	N/A
Oil capacity	N/A
Cooling system capacity	9.0 litres

Make/model Citroen ZX D/TD 1.9

Year **1991-94**

Engine (code)	1905cc (XUD9/T)
Power	71bhp/92bhp
Compression pressure	28bar approx
Firing order	1342
No.1 cyl	@ gearbox end
Valve clearances	(cold)
• inlet	0.15mm
• exhaust	0.30mm
Inj. pump type	Bosch or Lucas
Idle speed	800rpm
Max. no-load speed	5100rpm
Glow plug voltage	11

Pre-heat duration	5-10secs approx
Engine oil viscosity	15W/40 or 50
Oil quality	API CD
Oil capacity (inc. filter)	5.0 litres
Gearbox oil grade	SAE 75W/80
Oil capacity	1.9 litres
Final drive oil grade	N/A
Oil capacity	N/A
Cooling system capacity	8.5 litres (9.0 litres -Turbo)

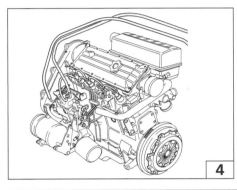

FIAT

Make/model Fiat Uno 60DS/Regata D

Year 1986-93

Engine (code)	1697cc (149.B3/146.B2))
Power	59bhp
Compression pressure	-
Firing order	1342
No.1 cyl	@ cambelt end
Valve clearances	(cold)
• inlet	0.30mm
• exhaust	0.30mm
Inj. pump type	Bosch
Idle speed	750rpm
Max. no-load speed	5100rpm
Glow plug voltage	11
Pre-heat duration	7-10secs approx
Engine oil viscosity	10W/30
Oil quality	API CD
Oil capacity (inc. filter)	4.0 litres
Gearbox oil grade	SAE 75W/80
Oil capacity	2.0 litres
Final drive oil grade	N/A
Oil capacity	N/A
Cooling system capacity	8.9 litres

Make/model Fiat Tipo/Tempra D/TD 1.9

Year 1990-94

Engine (code)	1929cc (160 A6/A7)
Power	65/91bhp
Compression pressure	-
Firing order	1342
No.1 cyl	@ cambelt end
Valve clearances	(cold)
• inlet	0.30mm
• exhaust	0.35mm
Inj. pump type	Bosch or Lucas
Idle speed	775rpm (900rpm -Turbo)
Max. no-load speed	5150rpm (5000rpm -Turbo)
Glow plug voltage	11
Pre-heat duration	6-10secs approx
Engine oil viscosity	10W/30
Oil quality	API CD
Oil capacity (inc. filter)	5.0 litres
Gearbox oil grade	SAE 80W
Oil capacity	1.4 litres
Final drive oil grade	N/A
Oil capacity	N/A
Cooling system capacity	8.8 litres

FORD

Make/model Ford Fiesta/Escort/Orion D 1.6

Year 1983-89

Engine (code)	1608cc (LT-)
Power	54bhp
Compression pressure	28-34bar
Firing order	1342
No.1 cyl	@ cambelt end
Valve clearances	(cold)
• inlet	0.24-0.47mm
• exhaust	0.44-0.57mm
Inj. pump type	Bosch or Lucas
Idle speed	875rpm

Max. no-load speed	4800rpm
Glow plug voltage	11.5
Pre-heat duration	8-10secs

Engine oil viscosity	10W/30
Oil quality	API CC or CD
Oil capacity (inc. filter)	5.0 litres
Gearbox oil grade	SAE 75W/80
Oil capacity	2.8 litres (3.1 litres -5-speed)
Final drive oil grade	N/A
Oil capacity	N/A

Cooling system capacity	9.3 litres (8.5 litres - Fiesta)

Make/model Ford Fiesta/Escort/Orion D 1.8

Year 1989-93

Engine (code)	1753cc (RT-)
Power	60bhp
Compression pressure	28-34bar
Firing order	1342
No.1 cyl	@ cambelt end
Valve clearances	(cold)
• inlet	0.30-0.40mm
• exhaust	0.45-0.55mm

Inj. pump type	Bosch or Lucas
Idle speed	850rpm
Max. no-load speed	5300-5400rpm
Glow plug voltage	11.5
Pre-heat duration	5secs

Engine oil viscosity	10W/30
Oil quality	API CC or CD
Oil capacity (inc. filter)	4.5 litres
Gearbox oil grade	SAE 80W/90
Oil capacity	3.1 litres
Final drive oil grade	N/A
Oil capacity	N/A

Cooling system capacity	9.3 litres (8.5 litres - Fiesta)

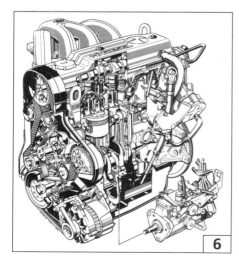

6

Make/model Ford Sierra/Sapphire TD 1.8

Year 1990-93

Engine (code)	1753cc (RF-)
Power	75bhp
Compression pressure	28-34bar
Firing order	1342
No.1 cyl	@ cambelt end
Valve clearances	(cold)
• inlet	0.30-0.40mm
• exhaust	0.45-0.55mm

Inj. pump type	Lucas
Idle speed	850rpm
Max. no-load speed	5100-5200rpm
Glow plug voltage	11.5
Pre-heat duration	5secs

Engine oil viscosity	10W/30
Oil quality	API CD
Oil capacity (inc. filter)	5.1 litres
Gearbox oil grade	SAE 75W/90
Oil capacity	1.25 litres
Final drive oil grade	90W
Oil capacity	0.9 litres

Cooling system capacity	9.5 litres

Make/model Ford Sierra/Sapphire D 2.3

Year 1983-89

Engine (code)	2304cc (YT-)
Power	67bhp
Compression pressure	20-25bar
Firing order	1342
No.1 cyl	@ gearbox end
Valve clearances	(cold)
• inlet	0.30mm
• exhaust	0.35mm

Inj. pump type	Lucas
Idle speed	800-850rpm
Max. no-load speed	4800rpm
Glow plug voltage	11.5
Pre-heat duration	7-11secs

Engine oil viscosity	15W/40
Oil quality	API CC or CD
Oil capacity (inc. filter)	5.6 litres
Gearbox oil grade	SAE 80W
Oil capacity	1.9 litres
Final drive oil grade	90W
Oil capacity	0.9 litres

Cooling system capacity	9.5 litres

PEUGEOT

Make/model Peugeot 205/305/309D 1.8/Turbo

Year 1983-94

Engine (code)	1769cc (XUD7/T)
Power	60bhp/78bhp
Compression pressure	25-30bar
Firing order	1342
No.1 cyl	@ gearbox end
Valve clearances	(cold)
• inlet	0.15mm
• exhaust	0.30mm
Inj. pump type	Bosch or Lucas
Idle speed	800rpm
Max. no-load speed	4600rpm (4700-4900 -Turbo)
Glow plug voltage	11
Pre-heat duration	7-10secs
Engine oil viscosity	10W/40 or 50
Oil quality	API CD
Oil capacity (inc. filter)	5.0 litres
Gearbox oil grade	SAE 75W/80 (BE3 box -10W/40)
Oil capacity	2.0 litres
Final drive oil grade	N/A
Oil capacity	N/A
Cooling system capacity	8.5 litres (7.8l -309Turbo, 9.5 litres -305D)

7

Make/model Peugeot 205/305/309D 1.9

Year 1982-94

Engine (code)	1905cc (XUD9)
Power	64bhp
Compression pressure	25-30bar
Firing order	1342
No.1 cyl	@ gearbox end
Valve clearances	(cold)

• inlet	0.15mm
• exhaust	0.30mm
Inj. pump type	Bosch or Lucas
Idle speed	800rpm
Max. no-load speed	4600rpm
Glow plug voltage	11
Pre-heat duration	7-10secs
Engine oil viscosity	10W/40 or 50
Oil quality	API CC or CD
Oil capacity (inc. filter)	5.0 litres
Gearbox oil grade	SAE 75W/80 (BE3 box -10W/40)
Oil capacity	2.0 litres
Final drive oil grade	N/A
Oil capacity	N/A
Cooling system capacity	8.5 litres (9.5 litres -305D)

Make/model Peugeot 405TD 1.8

Year 1988-94

Engine (code)	1769cc (XUD7TE)
Power	90bhp
Compression pressure	25-30bar
Firing order	1342
No.1 cyl	@ gearbox end
Valve clearances	(cold)
• inlet	0.15mm
• exhaust	0.30mm
Inj. pump type	Bosch or Lucas
Idle speed	750-800rpm
Max. no-load speed	4700-4900rpm
Glow plug voltage	11
Pre-heat duration	7-10secs
Engine oil viscosity	10W/40 or 50
Oil quality	API CD
Oil capacity (inc. filter)	5.0 litres
Gearbox oil grade	SAE 75W/80 (BE3 box -10W/40)
Oil capacity	2.0 litres
Final drive oil grade	N/A
Oil capacity	N/A
Cooling system capacity	7.8 litres

Make/model Peugeot 405D/TD

Year 1988-94

Engine (code)	1905cc (XUD9/TE)
Power	69/92bhp
Compression pressure	25-30bar
Firing order	1342
No.1 cyl	@ gearbox end

Valve clearances | (cold)
• inlet | 0.15mm
• exhaust | 0.30mm

Inj. pump type | Bosch or Lucas
Idle speed | 775rpm
Max. no-load speed | 5100rpm
Glow plug voltage | 11
Pre-heat duration | 7-10secs

Engine oil viscosity | 10W/40 or 50
 Oil quality | API CD
 Oil capacity (inc. filter) | 5.0 litres
Gearbox oil grade | SAE 75W/80
 | (BE3 box -10W/40)
 Oil capacity | 2.0 litres
Final drive oil grade | N/A
 Oil capacity | N/A

Cooling system capacity | 7.8 litres

RENAULT

Make/model Renault 5/9/11D 1.6

Year 1983-90

Engine (code) | 1596cc (F8M)
Power | 55bhp
Compression pressure | 20 bar min.
Firing order | 1342
No.1 cyl | @ gearbox end
Valve clearances | (cold)
• inlet | 0.20mm
• exhaust | 0.40mm

Inj. pump type | Bosch or Lucas
Idle speed | 850rpm
Max. no-load speed | 5250rpm
Glow plug voltage | 11
Pre-heat duration | 10-20secs

Engine oil viscosity | 10W/40
 Oil quality | API CC or CD
 Oil capacity (inc. filter) | 5.3 litres
Gearbox oil grade | SAE 80W
 Oil capacity | 3.4 litres
Final drive oil grade | N/A
 Oil capacity | N/A

Cooling system capacity | 6.7 litres (6.4 litres -R5)

Make/model Renault Clio/19/Chamade D/TD 1.9

Year 1989-94

Engine (code) | 1870cc (F8Q)
Power | 64/93bhp

Compression pressure | 20bar min.
Firing order | 1342
No.1 cyl | @ gearbox end
Valve clearances | (cold)
• inlet | 0.20mm
• exhaust | 0.40mm

Inj. pump type | Bosch or Lucas
Idle speed | 825rpm
Max. no-load speed | 5250rpm (4900rpm -Turbo)
Glow plug voltage | 11
Pre-heat duration | 10-20 secs

Engine oil viscosity | 15W/40
 Oil quality | API CD
 Oil capacity (inc. filter) | 5.5 litres
Gearbox oil grade | SAE 80W
 Oil capacity | 3.4 litres
Final drive oil grade | N/A
 Oil capacity | N/A

Cooling system capacity | 6.6 litres

ROVER

Make/model Rover Maestro/Montego D/TD 2.0

Year 1989-94

Engine (code) | 1994cc (MDi)
Power | 60/82bhp
Compression pressure | 35bar approx.
Firing order | 1342
No.1 cyl | @ cambelt end
Valve clearances | (cold)
• inlet | 0.25-0.35mm
• exhaust | 0.35-0.45mm

Inj. pump type | Bosch
Idle speed | 825rpm
Max. no-load speed | 5100rpm
Glow plug voltage | 11
Pre-heat duration | 4-10secs

Engine oil viscosity | 10W/40
 Oil quality | API CD
 Oil capacity (inc. filter) | 5.9 litres
Gearbox oil grade | 10W/30 or 40
 Oil capacity | 2.0 litres
Final drive oil grade | N/A
 Oil capacity | N/A

Cooling system capacity | 7.5 litres

Make/model Rover 218/418TD 1.8

Year 1991-94

Engine (code)	1769cc (XUD7TE)
Power	90bhp
Compression pressure	25-30bar
Firing order	1342
No.1 cyl	@ gearbox end
Valve clearances	(cold)
• inlet	0.15mm
• exhaust	0.30mm
Inj. pump type	Lucas
Idle speed	750-850rpm
Max. no-load speed	4300rpm
Glow plug voltage	11
Pre-heat duration	7-10secs
Engine oil viscosity	10W/30
Oil quality	API CD
Oil capacity (inc. filter)	4.5 litres
Gearbox oil grade	SAE 10W/40
Oil capacity	2.0 litres
Final drive oil grade	N/A
Oil capacity	N/A
Cooling system capacity	8.0 litres

Make/modelRover 218/418D 1.9

Year 1991-94

Engine (code)	1905cc (XUD9)
Power	70bhp
Compression pressure	25-30bar
Firing order	1342
No.1 cyl	@ gearbox end
Valve clearances	(cold)
• inlet	0.15mm
• exhaust	0.30mm
Inj. pump type	Lucas
Idle speed	750-850rpm
Max. no-load speed	4450-4750rpm
Glow plug voltage	11
Pre-heat duration	7-10secs
Engine oil viscosity	10W/30

Oil quality	API CC or CD
Oil capacity (inc. filter)	4.5 litres
Gearbox oil grade	SAE 10W/40
Oil capacity	2.0 litres
Final drive oil grade	N/A
Oil capacity	N/A
Cooling system capacity	8.0 litres

VAUXHALL

Make/model Vauxhall Nova D/TD 1.5

Year 1988-93

Engine (code)	1488cc (15D/T)
Power	50/67bhp
Compression pressure	22-25bar
Firing order	1342
No.1 cyl	@ cambelt end
Valve clearances	(cold)
• inlet	0.15mm
• exhaust	0.25mm
Inj. pump type	Bosch
Idle speed	825rpm
Max. no-load speed	5700rpm (5500 -Turbo)
Glow plug voltage	5
Pre-heat duration	8-15secs
Engine oil viscosity	10W/40
Oil quality	API CD
Oil capacity (inc. filter)	3.8 litres
Gearbox oil grade	SAE 80W
Oil capacity	1.8 litres
Final drive oil grade	N/A
Oil capacity	N/A
Cooling system capacity	6.5 litres

Make/model Vauxhall Astra/Cavalier D 1.6

Year 1982-89

Engine (code)	1598cc (16D)
Power	54bhp
Compression pressure	20bar approx.
Firing order	1342
No.1 cyl	@ cambelt end
Valve clearances	(hydraulic - no adjustment)
• inlet	-
• exhaust	-
Inj. pump type	Bosch
Idle speed	800-900rpm
Max. no-load speed	5000rpm

Glow plug voltage	11
Pre-heat duration	10-20secs
Engine oil viscosity	10W/40
Oil quality	API CC or CD
Oil capacity (inc. filter)	5.0 litres (3.75 litres to 84)
Gearbox oil grade	SAE 80W/90
Oil capacity	2.0 litres
Final drive oil grade	N/A
Oil capacity	N/A
Cooling system capacity	7.7 litres

M215 MUY 9

Make/model   Vauxhall Astra/Cavalier D 1.7

Year 1989-94

Engine (code)	1699cc (17D)
Power	57bhp
Compression pressure	20bar approx.
Firing order	1342
No.1 cyl	@ cambelt end
Valve clearances	(hydraulic - no adjustment)
• inlet	-
• exhaust	-
Inj. pump type	Bosch or Lucas
Idle speed	875rpm
Max. no-load speed	5500-5600rpm
Glow plug voltage	11
Pre-heat duration	10secs approx.
Engine oil viscosity	10W/40
Oil quality	API CD
Oil capacity (inc. filter)	4.75 litres
Gearbox oil grade	SAE 80W/90
Oil capacity	2.1 litres (1.7 litres -Cavalier)
Final drive oil grade	N/A
Oil capacity	N/A
Cooling system capacity	9.1 litres

Make/model Vauxhall Astra/Cavalier TD 1.7

Year 1990-94

Engine (code)	1686cc (17DT)
Power	82bhp
Compression pressure	-
Firing order	1342
No.1 cyl	@ cambelt end
Valve clearances	(cold)
• inlet	0.15mm
• exhaust	0.25mm
Inj. pump type	Bosch
Idle speed	780-880rpm
Max. no-load speed	5100-5300rpm
Glow plug voltage	11
Pre-heat duration	8-15secs
Engine oil viscosity	10W/40
Oil quality	API CD
Oil capacity (inc. filter)	4.5 litres
Gearbox oil grade	SAE 80W/90
Oil capacity	1.9 litres
Final drive oil grade	N/A
Oil capacity	N/A
Cooling system capacity	6.8 litres (7.4 litres -Cavalier)

L215 EVV 10

VOLKSWAGON

Make/model VW Golf & Jetta D 1.6

Year 1983-92

Engine (code)	1588cc (JP)
Power	55bhp
Compression pressure	26-34bar
Firing order	1342
No.1 cyl	@ cambelt end
Valve clearances	(hot) (hyd. - 86 on)
• inlet	0.25mm
• exhaust	0.45mm
Inj. pump type	Bosch or Lucas
Idle speed	850-900rpm
Max. no-load speed	5300-5400rpm
Glow plug voltage	11
Pre-heat duration	7-10secs approx
Engine oil viscosity	15W/40 or 50

Oil quality	**API CC or CD**
Oil capacity (inc. filter)	**3.5 litres (4.5 litres from 86)**
Gearbox oil grade	**SAE 80W**
Oil capacity	**1.5 litres (2.0 litres - 5speed)**
Final drive oil grade	**N/A**
Oil capacity	**N/A**
Cooling system capacity	**6.5 litres**

Make/model   VW Golf & Jetta D 1.6

Year `1983-92`

Engine (code)	**1588cc (JP)**
Power	**55bhp**
Compression pressure	**26-34bar**
Firing order	**1342**
No.1 cyl	**@ cambelt end**
Valve clearances	**(hot) (hyd. - 86 on)**
• inlet	**0.25mm**
• exhaust	**0.45mm**
Inj. pump type	**Bosch or Lucas**
Idle speed	**850-900rpm**
Max. no-load speed	**5300-5400rpm**
Glow plug voltage	**11**
Pre-heat duration	**7-10secs approx**
Engine oil viscosity	**15W/40 or 50**
Oil quality	**API CC or CD**
Oil capacity (inc. filter)	**3.5 litres (4.5 litres from 86)**
Gearbox oil grade	**SAE 80W**
Oil capacity	**1.5 litres (2.0 litres - 5speed)**
Final drive oil grade	**N/A**
Oil capacity	**N/A**
Cooling system capacity	**6.5 litres**

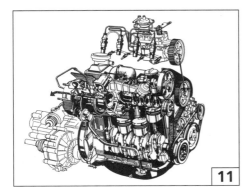

11

Make/model VW Golf & Jetta D 1.6

Year `1983-92`

Engine (code)	**1588cc (JP)**
Power	**55bhp**
Compression pressure	**26-34bar**
Firing order	**1342**
No.1 cyl	**@ cambelt end**
Valve clearances	**(hot) (hyd. - 86 on)**

• inlet	**0.25mm**
• exhaust	**0.45mm**
Inj. pump type	**Bosch or Lucas**
Idle speed	**850-900rpm**
Max. no-load speed	**5300-5400rpm**
Glow plug voltage	**11**
Pre-heat duration	**7-10secs approx**
Engine oil viscosity	**15W/40 or 50**
Oil quality	**API CC or CD**
Oil capacity (inc. filter)	**3.5 litres (4.5 litres from 86)**
Gearbox oil grade	**SAE 80W**
Oil capacity	**1.5 litres (2.0 litres - 5speed)**
Final drive oil grade	**N/A**
Oil capacity	**N/A**
Cooling system capacity	**6.5 litres**

Make/model VW Golf/Jetta TD 1.6

Year `1983-90`

Engine (code)	**1588cc (JR)**
Power	**70bhp**
Compression pressure	**26-34bar**
Firing order	**1342**
No.1 cyl	**@ cambelt end**
Valve clearances	**(hot) (hyd. - 86 on)**
• inlet	**0.20-0.25mm**
• exhaust	**0.40-0.45mm**
Inj. pump type	**Bosch or Lucas**
Idle speed	**850rpm**
Max. no-load speed	**5100rpm**
Glow plug voltage	**11**
Pre-heat duration	**7-10secs approx**
Engine oil viscosity	**15W/40 or 50**
Oil quality	**API CC or CD**
Oil capacity (inc. filter)	**3.5 litres (4.5 litres - 86 on)**
Gearbox oil grade	**SAE 80W**
Oil capacity	**2.0 litres**
Final drive oil grade	**N/A**
Oil capacity	**N/A**
Cooling system capacity	**6.5 litres**

Make/model VW Golf/Jetta TD 1.6

Year `1989-92`

Engine (code)	**1588cc (RA,SB)**
Power	**80bhp**
Compression pressure	**26-34bar**
Firing order	**1342**
No.1 cyl	**@ cambelt end**
Valve clearances	**(hydraulic - no adjustment)**

• inlet	N/A
• exhaust	N/A
Inj. pump type	Bosch
Idle speed	875-925rpm
Max. no-load speed	5000-5200rpm
Glow plug voltage	11
Pre-heat duration	7-10secs approx
Engine oil viscosity	15W/40 or 50
Oil quality	API CD
Oil capacity (inc. filter)	4.5 litres
Gearbox oil grade	SAE 80W
Oil capacity	2.0 litres
Final drive oil grade	N/A
Oil capacity	N/A
Cooling system capacity	6.5 litres

Make/model VW Golf/Vento D 1.9

Year 1991-94

Engine (code)	1896cc (1Y)
Power	64bhp
Compression pressure	26-34bar
Firing order	1342
No.1 cyl	@ cambelt end
Valve clearances	(hydraulic - no adjustment)
• inlet	N/A
• exhaust	N/A
Inj. pump type	Bosch
Idle speed	875-925rpm
Max. no-load speed	5100-5200rpm
Glow plug voltage	11
Pre-heat duration	7-10secs approx
Engine oil viscosity	10W/30
Oil quality	API CC or CD
Oil capacity (inc. filter)	4.5 litres
Gearbox oil grade	SAE 80W
Oil capacity	2.0 litres
Final drive oil grade	N/A
Oil capacity	N/A
Cooling system capacity	6.5 litres

Make/model VW Golf TD 1.9

Year 1991-94

Engine (code)	1896cc (AAZ)
Power	75bhp
Compression pressure	26-34bar
Firing order	1342
No.1 cyl	@ cambelt end
Valve clearances	(hydraulic - no adjustment)

• inlet	-
• exhaust	-
Inj. pump type	Bosch
Idle speed	875-925rpm
Glow plug voltage	11
Pre-heat duration	7-10secs approx
Engine oil viscosity	10W/30
Oil quality	API CD
Oil capacity (inc. filter)	4.5 litres
Gearbox oil grade	SAE 80W
Oil capacity	2.0 litres
Final drive oil grade	N/A
Oil capacity	N/A
Cooling system capacity	6.5 litres

YOUR CAR if not included above

Make/model ....

Year ....

Engine (code)	....
Power	....
Compression pressure	....
Firing order	....
Valve clearances	....
• inlet	....
• exhaust	....
Inj. pump type	....
Idle speed	....
Glow plug voltage	....
Pre-heat duration	....
Engine oil viscosity	....
Oil quality	....
Oil capacity (inc. filter)	....
Gearbox oil grade	....
Oil capacity	....
Final drive oil grade	....
Oil capacity	....
Cooling system capacity	....

CHAPTER 9 - TOOLS AND EQUIPMENT

Although this manual is chiefly concerned with maintenance of the diesel engine, the tools and equipment mentioned in this chapter cover all aspects of DIY maintenance around the car.

Basic maintenance on any car can be carried out using a fairly simple, relatively inexpensive tool kit. There is no need to spend a fortune all at once - most owners who do their own servicing acquire their tools over many years. However, there are some items you simply cannot do without in order to properly carry out the work necessary to keep your car on the road. Therefore, in the following lists, we have concentrated on those items which are likely to be valuable aids to maintaining

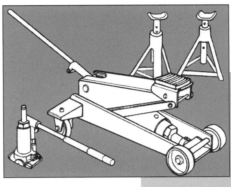

your car in a good state of tune, and to keep it running sweetly and safely. In addition we have featured some of the tools that are 'nice-to-have' rather than 'must have'. As your tool chest grows, there are some tools that help to make it easier to carry out your servicing and to be more thorough while you're at it. Much of the engine diagnostic and fuel injection system overhaul equipment we detail here is not necessary for everyday diesel maintenance, and is unrealistically expensive from the DIY point of view. However, we've made a point of mentioning and describing all the equipment that exists, in order to provide the enthusiast-mechanic with a thorough understanding of diesel workshop practice, and the less enthusiastic owner with some intuition of what he's paying for when he takes his diesel to a professional workshop!

Always buy the best quality tools you can afford. 'Cheap and cheerful' items may look similar to more expensive implements, but experience shows that they often fail when the going gets tough, and some can even be dangerous. With proper care, good quality tools will last a lifetime, and can be regarded as an investment. The extra outlay is well worth it, in the long run.

Diesel engines have many maintenance and repair requirements that differ from those of petrol engines. As such, some conventional equipment, particularly for tune-up or diagnostic work, is not suitable for the diesel. All, of course, will be revealed as you read on.

The following tool and equipment lists are shown under

Thanks are due to V.L. Churchill, Sykes Pickavant and Kamasa Tools for their kind assistance with this chapter. Many of the tools shown here were kindly supplied by these companies.

headings indicating the type of use applicable to each group.

Lifting:

SAFETY FIRST!
There are, of course, important safety implications when working underneath any vehicle. Sadly, DIY enthusiasts have been killed or seriously injured when maintaining their automotive pride and joy, usually for the want of a few moments thought. So - THINK SAFETY! In particular, NEVER venture beneath any vehicle supported only by a jack - of ANY type. A jack is ONLY intended to be a means of lifting a vehicle, NOT for holding it off the ground while being worked on. It is inevitable that you will need to raise the car from the ground in order to gain access to its underside. While the jack supplied with the car might be okay for emergency wheel-changes, it is not suitable for car servicing.

You need a good trolley jack, such as the 2 1/4 ton unit shown here (1.A). while alongside it is an excellent 'nice-to-have' - this extendible wheel nut spanner (1.B). This is also ideal for carrying in the car in case of punctures. If you've ever tried removing a wheel nut tightened by a garage gorilla, you know why this tool is so good!

Having raised the vehicle from the floor, always support it under a sound section of the 'chassis', or, if working at

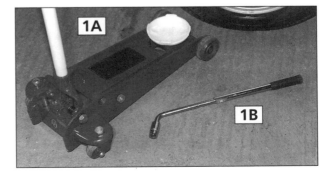

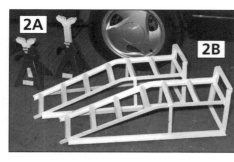

the rear of the car, beneath the rear axle. Use only proper axle stands (2.A), intended for the purpose, with solid wooden blocks on top, if necessary, to spread the load. These stands are exceptionally strong and are very rapidly adjusted, using the built-in ratchet stops. Screw-type stands are infinitely adjustable but are fiddly and time-consuming to use. NEVER, NEVER use bricks to support a car - they can crumble without warning, with horrifying results.

Always chock all wheels not in the air, and if possible, leave the car in first gear to prevent it from rolling.

Frankly, if you don't need to remove the road wheels for a particular job, the use of car ramps (2.B), which are generally more stable than axle stands - is preferable, in order to gain the necessary working height. However, there are dangers even with these. For further hints and tips concerning the use of car ramps, refer to *Chapter 1 - Safety First!*.

Jacks:

The manufacturer's jack is for emergency wheel changing ONLY - NOT to be used when working on the vehicle!

A 'bottle' jack - screw or hydraulic type - can be used as a means of lifting the car, in conjunction with axle stands to hold it clear of the ground. Ensure that the jack you buy is low enough to pass beneath your car's floorpan.

A trolley jack is extremely useful as it is so easily manoeuvrable. Again, use only for lifting the vehicle, in conjunction with axle stands to support it clear of the ground. Ensure that the lifting head of the jack will pass beneath the lowest points on the floorpan. Aim for the highest quality jack you can afford. Cheap types seldom last long, and can be VERY dangerous, suddenly allowing a car to drop to ground level - and proving the importance of a secondary support system (axle stands)!

Axle Stands:

Available in a range of sizes. Ensure that those you buy are sturdy, with the three legs reasonably widely spaced, and with a useful range of height adjustment.

Car Ramps:

Available in several heights - high ones are easier for working beneath the car, but too steep a ramp angle can cause problems if you have a low front spoiler, or exhaust.

The ultimate ramps are the 'wind-up' variety - easy to drive onto at their lowest height setting, then raised by means of

screw threads to a convenient working height.

Spanners:

Most diesel cars, being of European origin, and necessarily relatively modern in design, feature metric fasteners - nuts, bolts, studs and screws. Though there are some direct imperial equivalents (13/16 in. equals

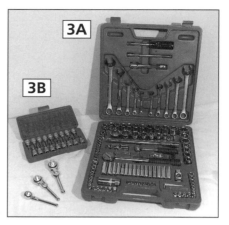

21 mm, for example) don't ever be tempted to use a non-metric spanner that is 'almost the right size'. Apart from the obvious danger of damaging the fastener beyond repair, there's also the very real possibility that you'll do the same to your hands!

This type of spanner set (3.A) is very useful in that it includes the more unusual types of spanner size in the same set. There are also 'stubby' ratchet handles available (3.B) for that cramped engine bay!

Note - in every case, ring spanners provide a more positive grip on a nut/bolt head than open-ended types, which can spread and/or slip when used on tight fasteners. Similarly, 'impact' type socket spanners with hexagonal apertures give better grip on a tight fastener than the normal 12 point 'bi-hex' variety.

Open-ended spanners can be bought in combined AF and metric sizes, and although this is an expensive way to equip yourself, it does cover every eventuality. You will need spanner sets covering as wide as possible a range - certainly from 3/8 to 15/16 in. for AF sizes, and 6 mm to 24 mm for metric.

Ring spanner sets should ideally cover much the same range to cover all types of work. Combination spanner sets (with one ring end, and one 'open' end of the same size, for each spanner) are particularly handy and versatile.

Socket spanners, again, should cover this range of sizes, and ideally both 3/8 in. and 1/2 in. square drive sets should be available - the former for small torque, difficult access situations, the latter for the opposite. A range of hex socket adapters should be included. A long extension bar is a typical 'nice-to-have' tool.

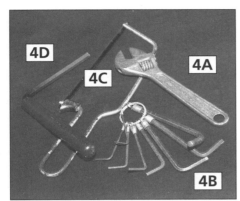

Buy yourself an adjustable spanner - nine inch, to start off with. (4.A)

Allen key set. (4.B)

A junior hacksaw (4.C)

Brake adjuster - essential for some models of disk brakes (4.D)

Universal or model-specific drain plug spanner. (5)

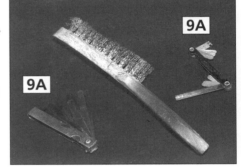

Torque wrench. This is very *nearly* a 'must-have' item and for any serious mechanic it *becomes* a 'must-have' once you have one. Prevents overtightening and stud shearing. (6)

Impact driver, useful for releasing seized screws. (7.A) Stud removing tools. A 'nice-to-have' when studs shear and all else fails (7.B). Hub pullers (7.C) are invariably needed for dismantling when you get beyond the simple servicing stage.

Screwdrivers:

General-purpose set of cross-head variety and a general purpose set of flat-bladed screwdrivers. (All available in various-sized sets.)

Sundry Items:

Tool box - steel types are sturdiest. Extension lead. Small/medium size ball pein hammer and

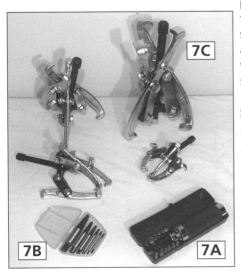

soft-faced hammer (8.B)

A set of drifts (8.C) is the sort of thing that you don't know how much you need until you've got one!

12 volt test lamp (can be made using 12 volt bulb, bulb holder, two short lengths of cable and two small crocodile clips).

Electrical test meter, including voltage, resistance and current (ammeter) functions - 'nice to have' for numerous electrical diagnoses, especially of the diesel's pre-heating system

A set of feeler gauges is no use on a diesel's spark plugs, of course, because there aren't any! But useful for checking other clearances and settings. (9.A)

'Automatic' valve clearance adjuster - 'nice to have' as an aid to accurate adjustment of screw-type rockers which have worn pads.

Copper-based anti-seize compound - useful during assembly of threaded components to make future dismantling easier!

Oil can (with multigrade engine oil, for general purpose lubrication).

Oil measuring jug. Oil filter strap or chain wrench. (10) Water dispellant 'electrical' aerosol spray. Pair of pliers ('standard' jaw).

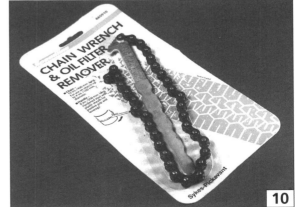

TOOLS AND EQUIPMENT

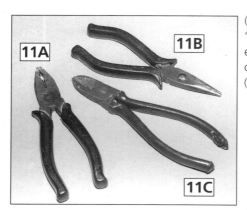

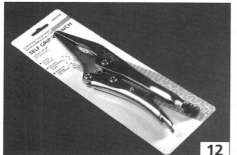

(11.A) Pair of 'long-nosed' pliers. (11.B) Pair of 'side cutters'. (11.C)

Self-grip wrench or - preferably - set of three. (12) Tyre pump. Tyre pressure gauge.

Tyre tread depth gauge.

Electric drill. Not a servicing tool as such but a 'must-have' nevertheless.

A rechargeable drill (13) is superb, enabling you to reach tight spots without trailing leads - and much safer out of doors. Recommended!

Diesel-Specific Equipment:

There is a number of companies that specialise in the manufacture or supply of diesel-specific workshop equipment, among them: A.M. Test Systems (Sykes-Pickavant), Bosch, V.L. Churchill, Dieselcare (Kent-Moore), Draper, Hartridge and Stanadyne. Bosch and Hartridge equipment is pretty exclusive stuff, mostly reserved for professionals, and while most diesel tooling is on the expensive side, and indeed targeted at professionals, there are some handy bits of equipment, particularly on offer by Dieseltune, Draper and Kent-Moore, that keen DIYers could well be tempted to purchase.

Depending on how deeply you wish to become involved with diesel maintenance and 'tune-up', you may end up needing the majority of the following items of equipment - see *Chapter 5, Diesel Servicing,* and *Chapter 6, Fault Finding,* for information on what each does and how to use them. The more expensive can be purchased gradually, as you save more and more money by doing your own jobs on the car! The good news is that none of the following items are actually

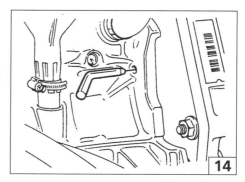

needed to accomplish routine diesel-car servicing. Most are related to fault diagnosis, injection timing or preventative mid-life maintenance.

14. For many types of engine, you will need a model-specific engine sprocket positioning pin, for checking and adjusting the injection timing (not needed for all diesel engines).

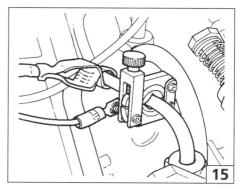

15. The popular piezo-electric point-of-injection pick-up clamps over the injector pipe of cylinder No. 1 to detect injection. This is used with timing lights, some test-rev counters, and multi-function diagnostic units, as used by the professionals. A diesel-specific timing light (looks like a petrol-engine timing light, but has a special pick-up on the end of its lead). You can also buy an adaptor unit to allow uses of conventional petrol-type engine timing lights or analysers. (Illustration, courtesy V. L. Churchill/LDV Limited)

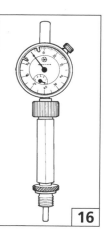

16. This is a dial gauge and injection pump-specific mounting kit. There are different sorts for different pumps. (Illustration, courtesy V. L. Churchill/LDV Limited)

17. Start-of-injection 'spill-pipe' - for determining injection timing on the less common in-line type of injection pump. (Illustration, courtesy Autodata)

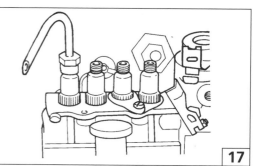

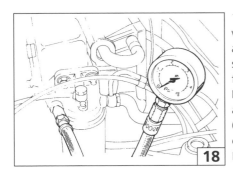

18. Pressure/ vacuum gauge and adaptors - for fuel system checks and fault diagnosis. Different connectors are also supplied. (Illustration, courtesy Ford Europe Ltd)

19. Glow plug testing device, if you really like pushing the boat out!. (Illustration, courtesy V L Churchill/LDV Limited)

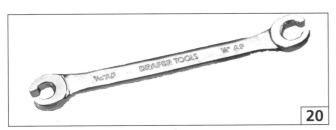

20. Specific-size split-ring spanner - for undoing injection pipe unions. Also known as a flare nut wrench. (Illustration, courtesy Ford Europe Ltd)

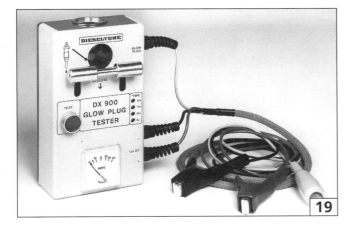

21. This is a 27 mm A/F injector spanner - for removal of the common indirect-injection (IDI) type of injector.

22. Specific-size injector extractor - for removal of the less common direct-injection (DI) type of injector.

The centre-piece is effectively a miniature slide-hammer, the surrounding items are adaptors for different injectors. (Illustration, courtesy V L Churchill/LDV Limited)

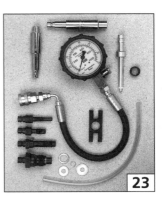

23. This diesel-specific cylinder compression gauge is for engine fault and condition diagnosis. Note the 'dummy' injector and glow plug adaptors that make it possible to test a wide variety of diesel engines. (Illustration, courtesy Sykes-Pickavant)

24. Diesel-specific rev-counter. This is V L Churchill/LDV Limited's 'optical' readout model. (Although how you read it if it isn't optical...)

25. Brass-bristle wire brush - for cleaning injector nozzles. Injector nozzle cleaning kit from Dieseltune includes brass brush, probes and scrapers (the latter for use on a dismantled nozzle). (Illustration, courtesy V L Churchill/LDV Limited)

26. A simple bench-mounted injector spray/pressure tester in use.

Note the test-fluid reservoir and pressure gauge. (Illustration, courtesy V L Churchill/LDV Limited)

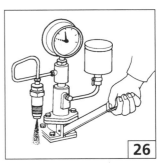

27. On-car injector spray and pressure tester. Dieseltune's DX710 injector tester is inexpensive and a great aid to injector fault diagnosis. (Illustration, courtesy V L Churchill/LDV Limited)

There's also such a thing as a multi-function diesel diagnostic unit - for checking dynamic injection timing, counting engine rpm, checking timing advance. It's far too pricey to be considered for DIY use, and you can carry out all these jobs with separates anyway. Definitely a 'lovely to have' though!

APPENDIX 1
RECOMMENDED CASTROL
LUBRICANTS

Castrol offers a broad range of top-quality oils, greases and other essential fluids, and they will all figure in the 'Recommended' list appended to the handbooks issued by the major motor manufacturers.

ENGINE OIL*

Castrol GTD Diesel

*Except in the following cases, where a CCMC D5 Lubricant is required:

Rover 825 Turbo Diesel 1990-on

Land Rover 110 & 90 Turbo Diesel 1987-1990

Range Rover Turbo Diesel 1986-1993

Castrol Turbomax

FLUSHING OIL

Castrol Solvent Flushing Oil

COOLANT

Castrol Anti-freeze & Summer Coolant

CASTROL ANTI-FREEZE: Recommended for use in petrol or diesel engine cooling systems, with aluminium or cast engines. Its formulation of mono ethylene glycol and corrosion inhibitors makes it suitable for all-year-round use, and because it contains no phosphate it is reckoned that the problems of deposits in some modern uprated engines are eliminated. A 33 per cent concentration will protect down to minus 17 degrees C.

WATER DISPLACEMENT FLUID

Castrol DWF

NUT & BOLT RELEASE

Castrol Easing Oil

ELECTRICAL CONNECTIONS

Castrol DWF

APPENDIX 2
SPECIALISTS & SUPPLIERS
FEATURED IN THIS BOOK

All of the products and specialists listed below have contributed in various ways to this book. All of the consumer products used are available through regular high street outlets or by mail or from specialist suppliers.

Robert Bosch Ltd., P O Box 948, Broadwater Park, North Orbital Road, Denham, Uxbridge, Middlesex, UB9 5HJ. Tel: 01895 834466

Castrol (UK) Ltd., Burmah House, Pipers Way, Swindon, Wiltshire, SN3 1RE. Tel: 01793 452222
Contact Castrol's Consumer Technical Department Help Line on the above number for advice on lubrication recommendations.

Diesel Car Magazine, Wessex Buildings, Somerton Business Park, Somerton, Somerset, TA11 6SB. Tel: 01458 274447

Dieseltune (V.L. Churchill Ltd.), P O Box 3, London Road, Daventry, Northants, NN11 4NF. Tel: 01327 704461

Dinol (GB) Ltd., Dinol House, 98 Ock Street, Abingdon, Oxford, OX14 5DH. Tel: 01235 530677
Suppliers of Dinitrol rust proofing fluids, and are equipped to carry out rustproofing on vehicles.

Ford Motor Company Ltd., Eagle Way, Warley, Brentwood, Essex, CM13 3BW. Tel: 01277 253000

Gunson Ltd, Coppen Road, Dagenham, Essex, RM8 1NU. Tel: 0181 984 8855.

Electrical and electronic engine tuning equipment.
Lucas Industries plc (Automotive), Brueton House, New Road, Solihull, West Midlands, B91 3TA. Tel: 0121 627 6000

NGK Spark Plugs (UK) Ltd, 7-8-9 Garrick Industrial Centre, Hendon, London, NW9 6AQ. Tel: 0181 202 2151.

Top quality spark plugs.
Peugeot Talbot Motor Company plc, Aldermoor House, Aldermoor Lane, Coventry, CV3 1LT. Tel: 01203 884000

Rover Group Ltd., International House, Bickenhill, Birmingham, B37 7HQ. Tel: 0121 782 8000

SP Tyres UK Ltd, Fort Dunlop, Birmingham, B24 9QT. Tel: 0121 384 4444.

Manufacturers of Dunlop tyres.
V.A.G. (UK) Ltd., Yeomans Drive, Blakelands, Milton Keynes, Buckinghamshire, MK14 5AN. Tel: 01908 601777

Vauxhall Motors Ltd., Griffin House, Osborne Road, Luton, Bedfordshire, LU2 0SY. Tel: 01582 21122

Waste Oil Disposal
There are 1,300 listed waste oil disposal sites in the UK alone. PLEASE don't foul the environment by tipping waste oil into the drains or the ground. Find your nearest oil disposal point by running the National Rivers Authority on *FREEPHONE 0800 663366.*

APPENDIX 3
SERVICE HISTORY

Although this book is about your car's diesel engine, you will undoubtedly want to service the whole car and not just the engine, when service time comes around.

For that reason, you'll find that this Appendix contains a comprehensive Service Schedule covering the diesel engine, and then another, contained in italics, covering the rest of the vehicle.

If you'd like more, specific information, Porter Manuals have published a range of model-specific Service Guides as well as one for those with little or no experience, called _Absolute Beginners' Service Guide_. See page 2 for a list of titles currently available.

This Chapter helps you keep track of all the servicing carried out both on your diesel engine and on your vehicle as a whole - and it can even save you money! A vehicle with a Service History is always worth more than one without, and you can make full use of this section, even if you have a garage or mechanic carry out the work for you. It enables you to specify the jobs you want to have carried out to your vehicle and, of course, it enables you to keep that all-important Service History. And even if your vehicle doesn't have a 'history' going back to when it was new, keeping this Chapter complete will add to your vehicle's value when you come to sell it. Mind you, it obviously won't be enough to just tick the boxes: keep all your receipts when you buy oil, filters and other consumables or parts. That way, you'll also be able to return any faulty parts if needs be.

As you move through the Service Intervals you will notice that the work carried out at earlier intervals, is repeated at each later Service Interval. Every time a Job or set of Jobs is 'repeated' from an earlier Interval, we show it in a tinted area on the page.

You will also find that all the major Intervals, right up to the 'longest', contain Jobs that are unique to that Service Interval. That's why we have continued this Service History right up to the _48,000 Miles or every Three Years_ interval. So now, you will be able to service your car and keep a full record of the work, in the knowledge that your car has been looked after as well as anyone could wish for!

The Job Lists

Wherever possible, the Jobs listed in this section have been placed in a logical order or placed into groups that will help you make progress on the car. We have tried to save you too much in the way of unnecessary movement by grouping jobs around areas of the car. Therefore, at each Service Interval you will see the work grouped into Jobs that need carrying out in the Engine Bay, Around The Car or Under The Car.

You'll also see space at each Service Interval for you to write down the date, price and seller's name every time you buy consumables or accessories. And once again, do remember to keep your receipts! There's also space for you to date and sign the Service Record or for a garage's stamp to be applied.

Important Note!

The Service Jobs listed here are intended as a check list and a means of keeping a record of your vehicle's service history, not as a set of instructions for working on your car. It is most important that you refer to _Chapter 5, Diesel Servicing_ for full details of how to carry out each Job listed here and for essential _SAFETY_ information and, see also, _Chapter 1, Safety First!._

PART I: DIESEL ENGINE SERVICING

EVERY 500 MILES, WEEKLY OR BEFORE A LONG JOURNEY

This list is shown, complete, only once. It would have been a bit much to have provided the list 52 times over for use once a week throughout the year! Each job is, however, included with every longer Service list from 3,000 miles/Three Months-on so that each of the 'weekly' Jobs is carried out as part of every service.

- [ ] Job 1. Engine oil level.
- [ ] Job 2. Coolant level.
- [ ] Job 3. Battery electrolyte.

EVERY 3,000 MILES - OR EVERY THREE MONTHS, whichever comes first

All the Service Jobs in the tinted area have been carried forward from earlier service intervals and are to be repeated at this service.

First carry out all Jobs listed under earlier Service Intervals as applicable.

- [ ] Job 1. Engine oil level.
- [ ] Job 2. Coolant level.
- [ ] Job 3. Battery electrolyte.

- [ ] Job 4. Generator drive belt.
- [ ] Job 5. Pipes and hoses.
- [ ] Job 6. Inspect for leaks.

EVERY 6,000 MILES - OR EVERY SIX MONTHS, whichever comes first

All the Service Jobs in the tinted area have been carried forward from earlier service intervals and are to be repeated at this service.

First carry out all Jobs listed under earlier Service Intervals as applicable.

- [ ] Job 3. Battery electrolyte.
- [ ] Job 4. Generator drive belt.
- [ ] Job 5. Pipes and hoses.
- [ ] Job 6. Inspect for leaks.

- [ ] Job 7. Draining water from fuel filter.
- [ ] Job 8. Accelerator controls.
- [ ] Job 9. Coolant system check.
- [ ] Job 10. Coolant check.
- [ ] Job 11. Check water pump.
- [ ] Job 12. Check power steering fluid.
- [ ] Job 13. Check clutch adjustment.
- [ ] Job 14. Change engine oil.
- [ ] Job 15. Renew oil filter.

EVERY 9,000 MILES - OR EVERY NINE MONTHS, whichever comes first

All the Jobs at this Service Interval have been carried forward from earlier Service Intervals and are to be repeated at this service.

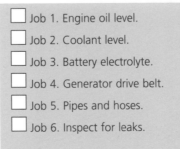

- [ ] Job 1. Engine oil level.
- [ ] Job 2. Coolant level.
- [ ] Job 3. Battery electrolyte.
- [ ] Job 4. Generator drive belt.
- [ ] Job 5. Pipes and hoses.
- [ ] Job 6. Inspect for leaks.

EVERY 12,000 MILES - OR EVERY TWELVE MONTHS, whichever comes first

All the Service Jobs in the tinted area have been carried forward from earlier service intervals and are to be repeated at this service.

First carry out all Jobs listed under earlier Service Intervals as applicable.

- [ ] Job 3. Battery electrolyte.
- [ ] Job 4. Generator drive belt.
- [ ] Job 5. Pipes and hoses.
- [ ] Job 6. Inspect for leaks.
- [ ] Job 8. Accelerator controls.
- [ ] Job 9. Coolant system check.
- [ ] Job 10. Coolant check.
- [ ] Job 11. Check water pump.
- [ ] Job 12. Check power steering fluid.
- [ ] Job 13. Check clutch adjustment.
- [ ] Job 14. Change engine oil.
- [ ] Job 15. Renew oil filter.

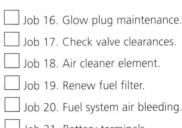

- [ ] Job 16. Glow plug maintenance.
- [ ] Job 17. Check valve clearances.
- [ ] Job 18. Air cleaner element.
- [ ] Job 19. Renew fuel filter.
- [ ] Job 20. Fuel system air bleeding.
- [ ] Job 21. Battery terminals.

EVERY 15,000 MILES - OR EVERY 15 MONTHS, whichever comes first

All the Jobs at this Service Interval have been carried forward from earlier Service Intervals and are to be repeated at this service.

☐ Job 1. Engine oil level.
☐ Job 2. Coolant level.
☐ Job 3. Battery electrolyte.
☐ Job 4. Generator drive belt.
☐ Job 5. Pipes and hoses.
☐ Job 6. Inspect for leaks.

EVERY 18,000 MILES - OR EVERY EIGHTEEN MONTHS, whichever comes first

All the Jobs at this Service Interval have been carried forward from earlier Service Intervals and are to be repeated at this service.

☐ Job 3. Battery electrolyte.
☐ Job 4. Generator drive belt.
☐ Job 5. Pipes and hoses.
☐ Job 6. Inspect for leaks.
☐ Job 7. Draining water from fuel filter.
☐ Job 8. Accelerator controls.
☐ Job 9. Coolant system check.
☐ Job 10. Coolant check.
☐ Job 11. Check water pump.
☐ Job 12. Check power steering fluid.
☐ Job 13. Check clutch adjustment.
☐ Job 14. Change engine oil.
☐ Job 15. Renew oil filter.

EVERY 21,000 MILES - OR EVERY TWENTY ONE MONTHS, whichever comes first

All the Jobs at this Service Interval have been carried forward from earlier Service Intervals and are to be repeated at this service.

☐ Job 1. Engine oil level.
☐ Job 2. Coolant level.
☐ Job 3. Battery electrolyte.
☐ Job 4. Generator drive belt.
☐ Job 5. Pipes and hoses.
☐ Job 6. Inspect for leaks.

EVERY 24,000 MILES - OR EVERY TWO YEARS, whichever comes first

All the Service Jobs in the tinted area have been carried forward from earlier service intervals and are to be repeated at this service.

First carry out all Jobs listed under earlier Service Intervals as applicable.

☐ Job 3. Battery electrolyte.
☐ Job 4. Generator drive belt.
☐ Job 5. Pipes and hoses.
☐ Job 6. Inspect for leaks.
☐ Job 8. Accelerator controls.
☐ Job 9. Coolant system check.
☐ Job 10. Coolant check.
☐ Job 11. Check water pump.
☐ Job 12. Check power steering fluid.
☐ Job 13. Check clutch adjustment.
☐ Job 14. Change engine oil.
☐ Job 15. Renew oil filter.
☐ Job 16. Glow plug maintenance.
☐ Job 17. Check valve clearances.
☐ Job 18. Air cleaner element.
☐ Job 19. Renew fuel filter.
☐ Job 20. Fuel system air bleeding.
☐ Job 21. Battery terminals.

☐ Job 22. Coolant renewal.
☐ Job 23. Checking/adjusting idle/fast idle speed.

EVERY 27,000 MILES - OR EVERY TWENTY SEVEN MONTHS, whichever comes first

All the Jobs at this Service Interval have been carried forward from earlier Service Intervals and are to be repeated at this service.

- ☐ Job 1. Engine oil level.
- ☐ Job 2. Coolant level.
- ☐ Job 3. Battery electrolyte.
- ☐ Job 4. Generator drive belt.
- ☐ Job 5. Pipes and hoses.
- ☐ Job 6. Inspect for leaks.

EVERY 33,000 MILES - OR EVERY THIRTY THREE MONTHS, whichever comes first

All the Jobs at this Service Interval have been carried forward from earlier Intervals and are to be repeated at this service.

- ☐ Job 1. Engine oil level.
- ☐ Job 2. Coolant level.
- ☐ Job 3. Battery electrolyte.
- ☐ Job 4. Generator drive belt.
- ☐ Job 5. Pipes and hoses.
- ☐ Job 6. Inspect for leaks.

EVERY 36,000 MILES - OR EVERY THREE YEARS, whichever comes first

All the Service Jobs in the tinted area have been carried forward from earlier service intervals and are to be repeated at this service.

First carry out all Jobs listed under earlier Service Intervals as applicable.

- ☐ Job 3. Battery electrolyte.
- ☐ Job 4. Generator drive belt.
- ☐ Job 5. Pipes and hoses.
- ☐ Job 6. Inspect for leaks.
- ☐ Job 8. Accelerator controls.
- ☐ Job 9. Coolant system check.
- ☐ Job 10. Coolant check.
- ☐ Job 11. Check water pump.
- ☐ Job 12. Check power steering fluid.
- ☐ Job 13. Check clutch adjustment.
- ☐ Job 14. Change engine oil.
- ☐ Job 15. Renew oil filter.
- ☐ Job 16. Glow plug maintenance.
- ☐ Job 17. Check valve clearances.
- ☐ Job 18. Air cleaner element.
- ☐ Job 19. Renew fuel filter.
- ☐ Job 20. Fuel system air bleeding.
- ☐ Job 21. Battery terminals.

☐ Job 24. Checking/adjusting injection timing.

☐ Job 25. Timing belt inspection/renewal.

EVERY 30,000 MILES - OR EVERY THIRTY MONTHS, whichever comes first

All the Jobs at this Service Interval have been carried forward from earlier Service Intervals and are to be repeated at this service.

- ☐ Job 3. Battery electrolyte.
- ☐ Job 4. Generator drive belt.
- ☐ Job 5. Pipes and hoses.
- ☐ Job 6. Inspect for leaks.
- ☐ Job 7. Draining water from fuel filter.
- ☐ Job 8. Accelerator controls.
- ☐ Job 9. Coolant system check.
- ☐ Job 10. Coolant check.
- ☐ Job 11. Check water pump.
- ☐ Job 12. Check power steering fluid.
- ☐ Job 13. Check clutch adjustment.
- ☐ Job 14. Change engine oil.
- ☐ Job 15. Renew oil filter.

EVERY 39,000 MILES - OR EVERY THIRTY NINE MONTHS, whichever comes first

All the Jobs at this Service Interval have been carried forward from earlier Service Intervals and are to be repeated at this service.

- ☐ Job 1. Engine oil level.
- ☐ Job 2. Coolant level.
- ☐ Job 3. Battery electrolyte.
- ☐ Job 4. Generator drive belt.
- ☐ Job 5. Pipes and hoses.
- ☐ Job 6. Inspect for leaks.

EVERY 42,000 MILES - OR EVERY FORTY TWO MONTHS, whichever comes first

All the Jobs at this Service Interval have been carried forward from earlier Service Intervals and are to be repeated at this service.

- ☐ Job 3. Battery electrolyte.
- ☐ Job 4. Generator drive belt.
- ☐ Job 5. Pipes and hoses.
- ☐ Job 6. Inspect for leaks.
- ☐ Job 7. Draining water from fuel filter.
- ☐ Job 8. Accelerator controls.
- ☐ Job 9. Coolant system check.
- ☐ Job 10. Coolant check.
- ☐ Job 11. Check water pump.
- ☐ Job 12. Check power steering fluid.
- ☐ Job 13. Check clutch adjustment.
- ☐ Job 14. Change engine oil.
- ☐ Job 15. Renew oil filter.

EVERY 45,000 MILES - OR EVERY FORTY FIVE MONTHS, whichever comes first

All the Jobs at this Service Interval have been carried forward from earlier Service Intervals and are to be repeated at this service.

- ☐ Job 1. Engine oil level.
- ☐ Job 2. Coolant level.
- ☐ Job 3. Battery electrolyte.
- ☐ Job 4. Generator drive belt.
- ☐ Job 5. Pipes and hoses.
- ☐ Job 6. Inspect for leaks.

EVERY 48,000 MILES - OR EVERY FOUR YEARS, whichever comes first

All the Service Jobs in the tinted area have been carried forward from earlier service intervals and are to be repeated at this service.

First carry out all Jobs listed under earlier Service Intervals as applicable.

- ☐ Job 3. Battery electrolyte.
- ☐ Job 4. Generator drive belt.
- ☐ Job 5. Pipes and hoses.
- ☐ Job 6. Inspect for leaks.
- ☐ Job 8. Accelerator controls.
- ☐ Job 9. Coolant system check.
- ☐ Job 11. Check water pump.
- ☐ Job 12. Check power steering fluid.
- ☐ Job 13. Check clutch adjustment.
- ☐ Job 14. Change engine oil.
- ☐ Job 15. Renew oil filter.
- ☐ Job 17. Check valve clearances.
- ☐ Job 18. Air cleaner element.
- ☐ Job 19. Renew fuel filter.
- ☐ Job 20. Fuel system air bleeding.
- ☐ Job 21. Battery terminals.
- ☐ Job 22. Coolant renewal.
- ☐ Job 23. Checking/adjusting idle/fast idle speed.

- ☐ Job 26. Glow plug renewal.
- ☐ Job 27. Injector maintenance.

PART II: THE REST OF THE CAR

EVERY 500 MILES, WEEKLY OR BEFORE A LONG JOURNEY

This list is shown, complete, only once. It would have been a bit much to have provided the list 52 times over for use once a week throughout the year! Each job is, however, included with every longer Service list from 3,000 miles/Three Months-on so that each of the 'weekly' Jobs is carried out as part of every service.

EVERY 500 MILES - THE ENGINE BAY

☐ Job 1. Brake fluid level.
☐ Job 2. Screenwash level.

EVERY 500 MILES - AROUND THE CAR

☐ Job 3. Check tyre pressures.
☐ Job 4. Check headlights, sidelights and front indicators.
☐ Job 5. Rear lights and indicators.
☐ Job 6. Interior lights.
☐ Job 7. Number plate lights.
☐ Job 8. Side repeater light bulbs.
☐ Job 9. Check horns.
☐ Job 10. Check windscreen wipers.
☐ Job 11. Windscreen washers.

EVERY 1,500 MILES - OR EVERY MONTH, whichever comes first

These Jobs are similar to the 500 Mile Jobs but don't need carrying out quite so regularly. Once again, these Jobs are not shown with a separate listing for each 1,500 miles/1 Month interval but they are included as part of every 3,000 miles/Three Months Service list and for every longer Service Interval.

EVERY 1,500 MILES - AROUND THE CAR

☐ Job 12. Check tyres.
☐ Job 13. Check spare tyre.
☐ Job 14. Wash bodywork.
☐ Job 15. Touch-up paintwork.
☐ Job 16. Aerial/Antenna.
☐ Job 17. Valet interior.
☐ Job 18. Improve visibility.

EVERY 1,500 MILES - UNDER THE CAR

☐ Job 19. Clean mud traps.

EVERY 3,000 MILES - OR EVERY THREE MONTHS, whichever comes first

All the Service Jobs in the tinted area have been carried forward from earlier service intervals and are to be repeated at this service.

EVERY 3,000 MILES - THE ENGINE BAY

First carry out all Jobs listed under earlier Service Intervals as applicable.

☐ Job 1. Brake fluid level.
☐ Job 2. Screenwash level.

EVERY 3,000 MILES - AROUND THE CAR

First carry out all Jobs listed under earlier Service Intervals as applicable.

- ☐ Job 3. Check tyre pressures.
- ☐ Job 4. Check headlights, side-lights and front indicators.
- ☐ Job 5. Rear lights and indicators.
- ☐ Job 6. Interior lights.
- ☐ Job 7. Number plate lights.
- ☐ Job 8. Side repeater light bulbs.
- ☐ Job 9. Check horns.
- ☐ Job 10. Check windscreen wipers.
- ☐ Job 11. Windscreen washers.
- ☐ Job 12. Check tyres.
- ☐ Job 13. Check spare tyre.
- ☐ Job 14. Wash bodywork.
- ☐ Job 15. Touch-up paintwork.
- ☐ Job 16. Aerial/Antenna.
- ☐ Job 17. Valet interior.
- ☐ Job 18. Improve visibility.

- ☐ Job 20. Check wheel bolts.
- ☐ Job 21. Check brake/fuel lines.
- ☐ Job 22. Check handbrake adjustment.
- ☐ Job 23. Door and tailgate seals.
- ☐ Job 24. Check windscreen.
- ☐ Job 25. Rear view mirrors.

EVERY 3,000 MILES - UNDER THE CAR

First carry out all Jobs listed under earlier Service Intervals as applicable.

- ☐ Job 19. Clean mud traps.

- ☐ Job 26. Check exhaust system.
- ☐ Job 27. Check exhaust mountings.
- ☐ Job 28. Check steering rack gaiters.
- ☐ Job 29. Check drive-shaft gaiters.
- ☐ Job 30. Steering joints.
- ☐ Job 31. Check suspension ball-joints.
- ☐ Job 32. Inspect for leaks.

EVERY 3,000 MILES - ROAD TEST

First carry out all Jobs listed under earlier Service Intervals as applicable.

- ☐ Job 33. Clean controls.
- ☐ Job 34. Check instruments.
- ☐ Job 35. Throttle pedal.
- ☐ Job 36. Brakes and steering.

EVERY 6,000 MILES - OR EVERY SIX MONTHS, whichever comes first

All the Service Jobs in the tinted area have been carried forward from earlier service intervals and are to be repeated at this service.

EVERY 6,000 MILES - THE ENGINE BAY

First carry out all Jobs listed under earlier Service Intervals as applicable.

- ☐ Job 1. Brake fluid level.
- ☐ Job 2. Screenwash level.

- ☐ Job 37. Radiator matrix.
- ☐ Job 38. Check manual gearbox oil.
- ☐ Job 39. Automatic transmission fluid.
- ☐ Job 40. Four wheel drive transmission.

EVERY 6,000 MILES - AROUND THE CAR

First carry out all Jobs listed under earlier Service Intervals as applicable.

☐ Job 3. Check tyre pressures.

☐ Job 4. Check headlights, side-lights and front indicators.

☐ Job 5. Rear lights and indicators.

☐ Job 6. Interior lights.

☐ Job 7. Number plate lights.

☐ Job 8. Side repeater light bulbs.

☐ Job 9. Check horns.

☐ Job 10. Check windscreen wipers.

☐ Job 11. Windscreen washers.

☐ Job 12. Check tyres.

☐ Job 13. Check spare tyre.

☐ Job 14. Wash bodywork.

☐ Job 15. Touch-up paintwork.

☐ Job 16. Aerial/Antenna.

☐ Job 17. Valet interior.

☐ Job 18. Improve visibility.

☐ Job 20. Check wheel bolts.

☐ Job 21. Check brake/fuel lines.

☐ Job 22. Check handbrake adjustment.

☐ Job 23. Door and tailgate seals.

☐ Job 24. Check windscreen.

☐ Job 25. Rear view mirrors.

☐ Job 41. Check seat belts.

☐ Job 42. Locks and hinges.

☐ Job 43. Bonnet release mechanism.

☐ Job 44. Check seats.

☐ Job 45. Test dampers.

☐ Job 46. Check/renew front disc brake pads.

☐ Job 47. 2.0 LITRE ENGINED MODELS ONLY Check/replace rear disc pads or shoes.

☐ Job 48. Check brake proportioning valve (if fitted).

EVERY 6,000 MILES - UNDER THE CAR

First carry out all Jobs listed under earlier Service Intervals as applicable.

☐ Job 19. Clean mud traps.

☐ Job 26. Check exhaust system.

☐ Job 27. Check exhaust mountings.

☐ Job 28. Check steering rack gaiters.

☐ Job 29. Check drive-shaft gaiters.

☐ Job 30. Steering joints.

☐ Job 31. Check suspension ball-joints.

☐ Job 32. Inspect for leaks.

EVERY 6,000 MILES - ROAD TEST

First carry out all Jobs listed under earlier Service Intervals as applicable.

☐ Job 33. Clean controls.

☐ Job 34. Check instruments.

☐ Job 35. Throttle pedal.

☐ Job 36. Brakes and steering.

EVERY 9,000 MILES - OR EVERY NINE MONTHS, whichever comes first

All the Jobs at this Service Interval have been carried forward from earlier Service Intervals and are to be repeated at this service.

EVERY 9,000 MILES - THE ENGINE BAY

☐ Job 1. Brake fluid level.

☐ Job 2. Screenwash level.

EVERY 9,000 MILES - AROUND THE CAR

☐ Job 3. Check tyre pressures.

☐ Job 4. Check headlights, side-lights and front indicators.

☐ Job 5. Rear lights and indicators.

☐ Job 6. Interior lights.

☐ Job 7. Number plate lights.

☐ Job 8. Side repeater light bulbs.

☐ Job 9. Check horns.

☐ Job 10. Check windscreen wipers.

☐ Job 11. Windscreen washers.

☐ Job 12. Check tyres.

☐ Job 13. Check spare tyre.

☐ Job 14. Wash bodywork.

☐ Job 15. Touch-up paintwork.

☐ Job 16. Aerial/Antenna.

☐ Job 17. Valet interior.

☐ Job 18. Improve visibility.

☐ Job 20. Check wheel bolts.

☐ Job 21. Check brake/fuel lines.

☐ Job 22. Check handbrake adjustment.

☐ Job 23. Door and tailgate seals.

☐ Job 24. Check windscreen.

☐ Job 25. Rear view mirrors.

EVERY 9,000 MILES - UNDER THE CAR

- ☐ Job 19. Clean mud traps.
- ☐ Job 26. Check exhaust system.
- ☐ Job 27. Check exhaust mountings.
- ☐ Job 28. Check steering rack gaiters.
- ☐ Job 29. Check drive-shaft gaiters.
- ☐ Job 30. Steering joints.
- ☐ Job 31. Check suspension ball-joints.
- ☐ Job 32. Inspect for leaks.

EVERY 9,000 MILES - ROAD TEST

- ☐ Job 33. Clean controls.
- ☐ Job 34. Check instruments.
- ☐ Job 35. Throttle pedal.
- ☐ Job 36. Brakes and steering.

EVERY 12,000 MILES - OR EVERY TWELVE MONTHS, whichever comes first

All the Service Jobs in the tinted area have been carried forward from earlier service intervals and are to be repeated at this service.

EVERY 12,000 MILES - THE ENGINE BAY

First carry out all Jobs listed under earlier Service Intervals as applicable.

- ☐ Job 2. Screenwash level.
- ☐ Job 37. Radiator matrix.
- ☐ Job 38. Check manual gearbox oil.
- ☐ Job 39. Automatic transmission fluid.
- ☐ Job 40. Four wheel drive transmission.

☐ Job 49. Read stored engine codes (if appropriate).

☐ Job 50. Check air conditioning.

EVERY 12,000 MILES - AROUND THE CAR

First carry out all Jobs listed under earlier Service Intervals as applicable.

- ☐ Job 3. Check tyre pressures.
- ☐ Job 4. Check headlights, side-lights and front indicators.
- ☐ Job 5. Rear lights and indicators.
- ☐ Job 6. Interior lights.
- ☐ Job 7. Number plate lights.
- ☐ Job 8. Side repeater light bulbs.
- ☐ Job 9. Check horns.
- ☐ Job 11. Windscreen washers.
- ☐ Job 12. Check tyres.
- ☐ Job 13. Check spare tyre.
- ☐ Job 14. Wash bodywork.
- ☐ Job 15. Touch-up paintwork.
- ☐ Job 16. Aerial/Antenna.
- ☐ Job 17. Valet interior.
- ☐ Job 18. Improve visibility.
- ☐ Job 20. Check wheel bolts.
- ☐ Job 21. Check brake/fuel lines.
- ☐ Job 22. Check handbrake adjustment.
- ☐ Job 23. Door and tailgate seals.
- ☐ Job 24. Check windscreen.
- ☐ Job 25. Rear view mirrors.
- ☐ Job 41. Check seat belts.
- ☐ Job 42. Locks and hinges.
- ☐ Job 43. Bonnet release mechanism.
- ☐ Job 44. Check seats.
- ☐ Job 45. Test dampers.
- ☐ Job 46. Check/renew front disc brake pads.
- ☐ Job 47. 2.0 LITRE ENGINED MODELS ONLY Check/replace rear disc pads or shoes.
- ☐ Job 48. Check brake proportioning valve (if fitted).

☐ Job 51. Toolkit and jack.

☐ Job 52. Light seals.

☐ Job 53. Alarm sender unit.

☐ Job 54. Adjust headlights.

☐ Job 55. Renew wiper blades.

☐ Job 56. Check hub bearings.

☐ Job 57. Check steering and suspension.

EVERY 12,000 MILES - UNDER THE CAR

First carry out all Jobs listed under earlier Service Intervals as applicable.

☐ Job 19. Clean mud traps.

☐ Job 26. Check exhaust system.

☐ Job 27. Check exhaust mountings.

☐ Job 28. Check steering rack gaiters.

☐ Job 29. Check drive-shaft gaiters.

☐ Job 30. Steering joints.

☐ Job 31. Check suspension ball-joints.

☐ Job 32. Inspect for leaks.

☐ Job 58. Inspect underside.

☐ Job 59. Clear drain holes.

☐ Job 60. Renew brake fluid.

EVERY 12,000 MILES - ROAD TEST

First carry out all Jobs listed under earlier Service Intervals as applicable.

☐ Job 33. Clean controls.

☐ Job 34. Check instruments.

☐ Job 35. Throttle pedal.

☐ Job 36. Brakes and steering.

EVERY 15,000 MILES - OR EVERY FIFTEEN MONTHS, whichever comes first

All the Jobs at this Service Interval have been carried forward from earlier Service Intervals and are to be repeated at this service.

EVERY 15,000 MILES - THE ENGINE BAY

☐ Job 1. Brake fluid level.

☐ Job 2. Screenwash level.

EVERY 15,000 MILES - AROUND THE CAR

☐ Job 3. Check tyre pressures.

☐ Job 4. Check headlights, side-lights and front indicators.

☐ Job 5. Rear lights and indicators.

☐ Job 6. Interior lights.

☐ Job 7. Number plate lights.

☐ Job 8. Side repeater light bulbs.

☐ Job 9. Check horns.

☐ Job 10. Check windscreen wipers.

☐ Job 11. Windscreen washers.

☐ Job 12. Check tyres.

☐ Job 13. Check spare tyre.

☐ Job 14. Wash bodywork.

☐ Job 15. Touch-up paintwork.

☐ Job 16. Aerial/Antenna.

☐ Job 17. Valet interior.

☐ Job 18. Improve visibility.

☐ Job 20. Check wheel bolts.

☐ Job 21. Check brake/fuel lines.

☐ Job 22. Check handbrake adjustment.

☐ Job 23. Door and tailgate seals.

☐ Job 24. Check windscreen.

☐ Job 25. Rear view mirrors.

EVERY 15,000 MILES - UNDER THE CAR

☐ Job 19. Clean mud traps.

☐ Job 26. Check exhaust system.

☐ Job 27. Check exhaust mountings.

☐ Job 28. Check steering rack gaiters.

☐ Job 29. Check drive-shaft gaiters.

☐ Job 30. Steering joints.

☐ Job 31. Check suspension ball-joints.

☐ Job 32. Inspect for leaks.

EVERY 15,000 MILES - ROAD TEST

☐ Job 33. Clean controls.

☐ Job 34. Check instruments.

☐ Job 35. Throttle pedal.

☐ Job 36. Brakes and steering.

EVERY 18,000 MILES - OR EVERY EIGHTEEN MONTHS, whichever comes first

All the Jobs at this Service Interval have been carried forward from earlier Service Intervals and are to be repeated at this service.

EVERY 18,000 MILES - THE ENGINE BAY

☐ Job 1. Brake fluid level.

☐ Job 2. Screenwash level.

☐ Job 37. Radiator matrix.

☐ Job 38. Check manual gearbox oil.

☐ Job 39. Automatic transmission fluid.

☐ Job 40. Four wheel drive transmission.

EVERY 18,000 MILES - AROUND THE CAR

☐ Job 3. Check tyre pressures.

☐ Job 4. Check headlights, side-lights and front indicators.

☐ Job 5. Rear lights and indicators.

☐ Job 6. Interior lights.

☐ Job 7. Number plate lights.

☐ Job 8. Side repeater light bulbs.

☐ Job 9. Check horns.

☐ Job 10. Check windscreen wipers.

☐ Job 11. Windscreen washers.

☐ Job 12. Check tyres.

☐ Job 13. Check spare tyre.

☐ Job 14. Wash bodywork.

☐ Job 15. Touch-up paintwork.

☐ Job 16. Aerial/Antenna.

☐ Job 17. Valet interior.

☐ Job 18. Improve visibility.

☐ Job 20. Check wheel bolts.

☐ Job 21. Check brake/fuel lines.

☐ Job 22. Check handbrake adjustment.

☐ Job 23. Door and tailgate seals.

☐ Job 24. Check windscreen.

☐ Job 25. Rear view mirrors.

☐ Job 41. Check seat belts.

☐ Job 42. Locks and hinges.

☐ Job 43. Bonnet release mechanism.

☐ Job 44. Check seats.

☐ Job 45. Test dampers.

☐ Job 46. Check/renew front disc brake pads.

☐ Job 47. 2.0 LITRE ENGINED MODELS ONLY Check/replace rear disc pads or shoes.

☐ Job 48. Check brake proportioning valve (if fitted).

EVERY 18,000 MILES - UNDER THE CAR

☐ Job 19. Clean mud traps.

☐ Job 26. Check exhaust system.

☐ Job 27. Check exhaust mountings.

☐ Job 28. Check steering rack gaiters.

☐ Job 29. Check drive-shaft gaiters.

☐ Job 30. Steering joints.

☐ Job 31. Check suspension ball-joints.

☐ Job 32. Inspect for leaks.

EVERY 18,000 MILES - ROAD TEST

☐ Job 33. Clean controls.

☐ Job 34. Check instruments.

☐ Job 35. Throttle pedal.

☐ Job 36. Brakes and steering.

EVERY 21,000 MILES - OR EVERY NINE MONTHS, whichever comes first

All the Jobs at this Service Interval have been carried forward from earlier Service Intervals and are to be repeated at this service.

EVERY 21,000 MILES - THE ENGINE BAY

- ☐ Job 1. Brake fluid level.
- ☐ Job 2. Screenwash level.

EVERY 21,000 MILES - AROUND THE CAR

- ☐ Job 3. Check tyre pressures.
- ☐ Job 4. Check headlights, side-lights and front indicators.
- ☐ Job 5. Rear lights and indicators.
- ☐ Job 6. Interior lights.
- ☐ Job 7. Number plate lights.
- ☐ Job 8. Side repeater light bulbs.
- ☐ Job 9. Check horns.
- ☐ Job 10. Check windscreen wipers.
- ☐ Job 11. Windscreen washers.
- ☐ Job 12. Check tyres.
- ☐ Job 13. Check spare tyre.
- ☐ Job 14. Wash bodywork.
- ☐ Job 15. Touch-up paintwork.
- ☐ Job 16. Aerial/Antenna.
- ☐ Job 17. Valet interior.
- ☐ Job 18. Improve visibility.
- ☐ Job 20. Check wheel bolts.
- ☐ Job 21. Check brake/fuel lines.
- ☐ Job 22. Check handbrake adjustment.
- ☐ Job 23. Door and tailgate seals.
- ☐ Job 24. Check windscreen.
- ☐ Job 25. Rear view mirrors.

EVERY 21,000 MILES - UNDER THE CAR

- ☐ Job 19. Clean mud traps.
- ☐ Job 26. Check exhaust system.
- ☐ Job 27. Check exhaust mountings.
- ☐ Job 28. Check steering rack gaiters.
- ☐ Job 29. Check drive-shaft gaiters.
- ☐ Job 30. Steering joints.
- ☐ Job 31. Check suspension ball-joints.
- ☐ Job 32. Inspect for leaks.

EVERY 21,000 MILES - ROAD TEST

- ☐ Job 33. Clean controls.
- ☐ Job 34. Check instruments.
- ☐ Job 35. Throttle pedal.
- ☐ Job 36. Brakes and steering.

EVERY 24,000 MILES - OR EVERY TWO YEARS, whichever comes first

All the Service Jobs in the tinted area have been carried forward from earlier service intervals and are to be repeated at this service.

EVERY 24,000 MILES - THE ENGINE BAY

First carry out all Jobs listed under earlier Service Intervals as applicable.

- ☐ Job 2. Screenwash level.
- ☐ Job 37. Radiator matrix.
- ☐ Job 38. Check manual gearbox oil.
- ☐ Job 39. Automatic transmission fluid.
- ☐ Job 40. Four wheel drive transmission.
- ☐ Job 49. Read stored engine codes (if appropriate).
- ☐ Job 50. Check air conditioning.

EVERY 24,000 MILES - AROUND THE CAR

First carry out all Jobs listed under earlier Service Intervals as applicable.

☐ Job 3. Check tyre pressures.

☐ Job 4. Check headlights, sidelights and front indicators.

☐ Job 5. Rear lights and indicators.

☐ Job 6. Interior lights.

☐ Job 7. Number plate lights.

☐ Job 8. Side repeater light bulbs.

☐ Job 9. Check horns.

☐ Job 11. Windscreen washers.

☐ Job 12. Check tyres.

☐ Job 13. Check spare tyre.

☐ Job 14. Wash bodywork.

☐ Job 15. Touch-up paintwork.

☐ Job 16. Aerial/Antenna.

☐ Job 17. Valet interior.

☐ Job 18. Improve visibility.

☐ Job 20. Check wheel bolts.

☐ Job 21. Check brake/fuel lines.

☐ Job 22. Check handbrake adjustment.

☐ Job 23. Door and tailgate seals.

☐ Job 24. Check windscreen.

☐ Job 25. Rear view mirrors.

☐ Job 41. Check seat belts.

☐ Job 42. Locks and hinges.

☐ Job 43. Bonnet release mechanism.

☐ Job 44. Check seats.

☐ Job 45. Test dampers.

☐ Job 46. Check/renew front disc brake pads.

☐ Job 47. 2.0 LITRE ENGINED MODELS ONLY Check/replace rear disc pads or shoes.

☐ Job 48. Check brake proportioning valve (if fitted).

☐ Job 51. Toolkit and jack.

☐ Job 52. Light seals.

☐ Job 53. Alarm sender unit.

☐ Job 54. Adjust headlights.

☐ Job 55. Renew wiper blades.

☐ Job 56. Check hub bearings.

☐ Job 57. Check steering and suspension.

☐ Job 61. Check brake discs/drums and calipers.

EVERY 24,000 MILES - UNDER THE CAR

First carry out all Jobs listed under earlier Service Intervals as applicable.

☐ Job 19. Clean mud traps.

☐ Job 26. Check exhaust system.

☐ Job 27. Check exhaust mountings.

☐ Job 28. Check steering rack gaiters.

☐ Job 29. Check drive-shaft gaiters.

☐ Job 30. Steering joints.

☐ Job 31. Check suspension ball-joints.

☐ Job 32. Inspect for leaks.

☐ Job 58. Inspect underside.

☐ Job 59. Clear drain holes.

☐ Job 60. Renew brake fluid.

EVERY 24,000 MILES - UNDER THE CAR

First carry out all Jobs listed under earlier Service Intervals as applicable.

☐ Job 33. Clean controls.

☐ Job 34. Check instruments.

☐ Job 35. Throttle pedal.

☐ Job 36. Brakes and steering.

EVERY 27,000 MILES - OR EVERY TWENTY SEVEN MONTHS, whichever comes first

All the Jobs at this Service Interval have been carried forward from earlier Service Intervals and are to be repeated at this service.

EVERY 27,000 MILES - THE ENGINE BAY

☐ Job 1. Brake fluid level.

☐ Job 2. Screenwash level.

EVERY 27,000 MILES - AROUND THE CAR

☐ Job 3. Check tyre pressures.

☐ Job 4. Check headlights, side-lights and front indicators.

☐ Job 5. Rear lights and indicators.

☐ Job 6. Interior lights.

☐ Job 7. Number plate lights.

☐ Job 8. Side repeater light bulbs.

☐ Job 9. Check horns.

☐ Job 10. Check windscreen wipers.

☐ Job 11. Windscreen washers.

☐ Job 12. Check tyres.

☐ Job 13. Check spare tyre.

☐ Job 14. Wash bodywork.

☐ Job 15. Touch-up paintwork.

☐ Job 16. Aerial/Antenna.

☐ Job 17. Valet interior.

☐ Job 18. Improve visibility.

☐ Job 20. Check wheel bolts.

☐ Job 21. Check brake/fuel lines.

☐ Job 22. Check handbrake adjustment.

☐ Job 23. Door and tailgate seals.

☐ Job 24. Check windscreen.

☐ Job 25. Rear view mirrors.

EVERY 27,000 MILES - UNDER THE CAR

- [ ] Job 19. Clean mud traps.
- [ ] Job 26. Check exhaust system.
- [ ] Job 27. Check exhaust mountings.
- [ ] Job 28. Check steering rack gaiters.
- [ ] Job 29. Check drive-shaft gaiters.
- [ ] Job 30. Steering joints.
- [ ] Job 31. Check suspension ball-joints.
- [ ] Job 32. Inspect for leaks.

EVERY 27,000 MILES - ROAD TEST

- [ ] Job 33. Clean controls.
- [ ] Job 34. Check instruments.
- [ ] Job 35. Throttle pedal.
- [ ] Job 36. Brakes and steering.

EVERY 30,000 MILES - OR EVERY THIRTY MONTHS, whichever comes first

All the Jobs at this Service Interval have been carried forward from earlier Service Intervals and are to be repeated at this service.

EVERY 30,000 MILES - THE ENGINE BAY

- [ ] Job 1. Brake fluid level.
- [ ] Job 2. Screenwash level.
- [ ] Job 37. Radiator matrix.
- [ ] Job 38. Check manual gearbox oil.
- [ ] Job 39. Automatic transmission fluid.
- [ ] Job 40. Four wheel drive transmission.

EVERY 30,000 MILES - AROUND THE CAR

- [ ] Job 3. Check tyre pressures.
- [ ] Job 4. Check headlights, side-lights and front indicators.
- [ ] Job 5. Rear lights and indicators.
- [ ] Job 6. Interior lights.
- [ ] Job 7. Number plate lights.
- [ ] Job 8. Side repeater light bulbs.
- [ ] Job 9. Check horns.
- [ ] Job 10. Check windscreen wipers.
- [ ] Job 11. Windscreen washers.
- [ ] Job 12. Check tyres.
- [ ] Job 13. Check spare tyre.
- [ ] Job 14. Wash bodywork.
- [ ] Job 15. Touch-up paintwork.
- [ ] Job 16. Aerial/Antenna.
- [ ] Job 17. Valet interior.
- [ ] Job 18. Improve visibility.
- [ ] Job 20. Check wheel bolts.
- [ ] Job 21. Check brake/fuel lines.
- [ ] Job 22. Check handbrake adjustment.
- [ ] Job 23. Door and tailgate seals.
- [ ] Job 24. Check windscreen.
- [ ] Job 25. Rear view mirrors.
- [ ] Job 41. Check seat belts.
- [ ] Job 42. Locks and hinges.
- [ ] Job 43. Bonnet release mechanism.
- [ ] Job 44. Check seats.
- [ ] Job 45. Test dampers.
- [ ] Job 46. Check/renew front disc brake pads.
- [ ] Job 47. 2.0 LITRE ENGINED MODELS ONLY Check/replace rear disc pads or shoes.
- [ ] Job 48. Check brake proportioning valve (if fitted).

EVERY 30,000 MILES - UNDER THE CAR

- ☐ Job 19. Clean mud traps.
- ☐ Job 26. Check exhaust system.
- ☐ Job 27. Check exhaust mountings.
- ☐ Job 28. Check steering rack gaiters.
- ☐ Job 29. Check drive-shaft gaiters.
- ☐ Job 30. Steering joints.
- ☐ Job 31. Check suspension ball-joints.
- ☐ Job 32. Inspect for leaks.

EVERY 30,000 MILES - ROAD TEST

- ☐ Job 33. Clean controls.
- ☐ Job 34. Check instruments.
- ☐ Job 35. Throttle pedal.
- ☐ Job 36. Brakes and steering.

EVERY 33,000 MILES - OR EVERY THIRTY THREE MONTHS, whichever comes first

All the Jobs at this Service Interval have been carried forward from earlier Service Intervals and are to be repeated at this service.

EVERY 33,000 MILES - UNDER THE CAR

- ☐ Job 1. Brake fluid level.
- ☐ Job 2. Screenwash level.

EVERY 33,000 MILES - UNDER THE CAR

- ☐ Job 3. Check tyre pressures.
- ☐ Job 4. Check headlights, side-lights and front indicators.
- ☐ Job 5. Rear lights and indicators.
- ☐ Job 6. Interior lights.
- ☐ Job 7. Number plate lights.
- ☐ Job 8. Side repeater light bulbs.
- ☐ Job 9. Check horns.
- ☐ Job 10. Check windscreen wipers.
- ☐ Job 11. Windscreen washers.
- ☐ Job 12. Check tyres.
- ☐ Job 13. Check spare tyre.
- ☐ Job 14. Wash bodywork.
- ☐ Job 15. Touch-up paintwork.
- ☐ Job 16. Aerial/Antenna.
- ☐ Job 17. Valet interior.
- ☐ Job 18. Improve visibility.
- ☐ Job 20. Check wheel bolts.
- ☐ Job 21. Check brake/fuel lines.
- ☐ Job 22. Check handbrake adjustment.
- ☐ Job 23. Door and tailgate seals.
- ☐ Job 24. Check windscreen.
- ☐ Job 25. Rear view mirrors.

EVERY 33,000 MILES - UNDER THE CAR

- ☐ Job 19. Clean mud traps.
- ☐ Job 26. Check exhaust system.
- ☐ Job 27. Check exhaust mountings.
- ☐ Job 28. Check steering rack gaiters.
- ☐ Job 29. Check drive-shaft gaiters.
- ☐ Job 30. Steering joints.
- ☐ Job 31. Check suspension ball-joints.
- ☐ Job 32. Inspect for leaks.

EVERY 33,000 MILES - ROAD TEST

- ☐ Job 33. Clean controls.
- ☐ Job 34. Check instruments.
- ☐ Job 35. Throttle pedal.
- ☐ Job 36. Brakes and steering.

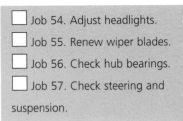

EVERY 36,000 MILES - OR EVERY THREE YEARS, whichever comes first

All the Service Jobs in the tinted area have been carried forward from earlier service intervals and are to be repeated at this service.

EVERY 36,000 MILES - THE ENGINE BAY

First carry out all Jobs listed under earlier Service Intervals as applicable.

☐ Job 2. Screenwash level.

☐ Job 37. Radiator matrix.

☐ Job 38. Check manual gearbox oil.

☐ Job 39. Automatic transmission fluid.

☐ Job 40. Four wheel drive transmission.

☐ Job 49. Read stored engine codes (if appropriate).

☐ Job 50. Check air conditioning.

EVERY 36,000 MILES - AROUND THE CAR

First carry out all Jobs listed under earlier Service Intervals as applicable.

☐ Job 3. Check tyre pressures.

☐ Job 4. Check headlights, side-lights and front indicators.

☐ Job 5. Rear lights and indicators.

☐ Job 6. Interior lights.

☐ Job 7. Number plate lights.

☐ Job 8. Side repeater light bulbs.

☐ Job 9. Check horns.

☐ Job 11. Windscreen washers.

☐ Job 12. Check tyres.

☐ Job 13. Check spare tyre.

☐ Job 14. Wash bodywork.

☐ Job 15. Touch-up paintwork.

☐ Job 16. Aerial/Antenna.

☐ Job 17. Valet interior.

☐ Job 18. Improve visibility.

☐ Job 20. Check wheel bolts.

☐ Job 21. Check brake/fuel lines.

☐ Job 22. Check handbrake adjustment.

☐ Job 23. Door and tailgate seals.

☐ Job 24. Check windscreen.

☐ Job 25. Rear view mirrors.

☐ Job 41. Check seat belts.

☐ Job 42. Locks and hinges.

☐ Job 43. Bonnet release mechanism.

☐ Job 44. Check seats.

☐ Job 45. Test dampers.

☐ Job 46. Check/renew front disc brake pads.

☐ Job 47. 2.0 LITRE ENGINED MODELS ONLY Check/replace rear disc pads or shoes.

☐ Job 48. Check brake proportioning valve (if fitted).

☐ Job 51. Toolkit and jack.

☐ Job 52. Light seals.

☐ Job 53. Alarm sender unit.

☐ Job 54. Adjust headlights.

☐ Job 55. Renew wiper blades.

☐ Job 56. Check hub bearings.

☐ Job 57. Check steering and suspension.

EVERY 36,000 MILES - UNDER THE CAR

First carry out all Jobs listed under earlier Service Intervals as applicable.

☐ Job 19. Clean mud traps.

☐ Job 26. Check exhaust system.

☐ Job 27. Check exhaust mountings.

☐ Job 28. Check steering rack gaiters.

☐ Job 29. Check drive-shaft gaiters.

☐ Job 30. Steering joints.

☐ Job 31. Check suspension ball-joints.

☐ Job 32. Inspect for leaks.

☐ Job 58. Inspect underside.

☐ Job 59. Clear drain holes.

☐ Job 60. Renew brake fluid.

☐ Job 62. Change transmission fluid.

☐ Job 63. Rustproofing.

EVERY 36,000 MILES - ROAD TEST

First carry out all Jobs listed under earlier Service Intervals as applicable.

☐ Job 33. Clean controls.

☐ Job 34. Check instruments.

☐ Job 35. Throttle pedal.

☐ Job 36. Brakes and steering.

AMERICAN AND BRITISH TERMS

It was Mark Twain who described the British and the Americans as, "two nations divided by a common language". such cynicism has no place here but we do acknowledge that our common language evolves in different directions. We hope that this glossary of terms, commonly encountered when servicing your car, will be of assistance to American owners and, in some cases, English speaking owners in other parts of the world, too.

American	British
Antenna	Antenna
Axleshaft	Halfshaft
Back-up	Reverse
Carburetor	Carburettor
Cotter pin	Split pin
Damper	Shock absorber
DC Generator	Dynamo
Defog	Demist
Drive line	Transmission
Driveshaft	Propeller shaft
Fender	Wing or mudguard
Firewall	Bulkhead
First gear	Bottom gear
Float bowl	Float chamber
Freeway, turnpike	Motorway
Frozen	Seized
Gas tank	Petrol tank
Gas pedal	Accelerator or throttle pedal
Gasoline, Gas or Fuel	Petrol or fuel
Ground (electricity)	Earth
Hard top	Fast back
Header	Exhaust manifold
Headlight dimmer	Headlamp dipswitch
High gear	Top gear
Hood	Bonnet
Industrial Alcohol or Denatured Alcohol	Methylated spirit
Kerosene	Paraffin
Lash	Free-play
License plate	Number plate
Lug nut	Wheel nut
Mineral spirit	White spirit
Muffler	Silencer
Oil pan	Sump
Panel wagon/van	Van
Parking light	Side light
Parking brake	Hand brake
'Pinging'	'Pinking'
Quarter window	Quarterlight
Recap (tire)	Remould or retread
Rocker panel	Sill panel

American	British
Rotor or disk (brake)	Disc
Sedan	Saloon
Sheet metal	Bodywork
Shift lever	Gear lever
Side marker lights, side turn signal or position indicator	Side indicator lights
Soft-top	Hood
Spindle arm	Steering arm
Stabiliser or sway bar	Anti-roll bar
Throw-out bearing	Release or thrust bearing
Tie-rod (or connecting rod)	Track rod (or steering)
Tire	Tyre
Transmission	Drive line
Trouble shooting	Fault finding/diagnosis
Trunk	Boot
Turn signal	Indicator
Valve lifter	Tappet
Valve cover	Rocker cover
Valve lifter or tappet	Cam follower or tappet
Vise	Vice
Windshield	Windscreen
Wrench	Spanner

Useful conversions:

	Multiply by
US gallons to Litres	3.785
Litres to US gallons	0.2642
UK gallons to US gallons	1.20095
US gallons to UK gallons	0.832674

Fahrenheit to Celsius (Centigrade) -
Subtract 32, multiply by 0.5555

Celsius to Fahrenheit -
Multiply by 1.8, add 32